Confessions of a Psychologist

# HUMAN INCOMPETENCE

## CONFESSIONS OF A
## PSYCHOLOGIST

Performance Management Publications (PMP)

Performance Management Publications (PMP)
3344 Peachtree Road NE, Suite 1050
Atlanta, GA 30326
678.904.6140
www.PManagementPubs.com

ISBN-13: 978-0-937100-21-9
ISBN-10: 0-937100-21-8

2 3 4 5 6 7

Cover Design: Lisa Smith
Text Design: James Omedo
Editor: Gail Snyder
Executive Editor: Darnell Lattal
Production Coordinator: Laura-Lee Glass

PMP books are available at special discounts for bulk purchases by corporations, institutions, and other organizations. For more information, please call 678.904.6140, ext. 131 or e-mail info@aubreydaniels.com.

Copyright © 2011 by Aubrey Daniels International, Inc.
All rights reserved. No part of this book may be reproduced in any form, electronic or mechanical, including photocopy, recording, or any information storage and retrieval system, without permission in writing from the publisher.

# HUMAN INCOMPETENCE

## CONFESSIONS OF A PSYCHOLOGIST

Confessions of a Psychologist

# CONTENTS
HUMAN INCOMPETENCE | Confessions of a Psychologist

*Preface to the work* . . . . . . . . . . . . . . . . . . . . . . . . *i*
*Foreword* . . . . . . . . . . . . . . . . . . . . . . . . . . . . . . . *xi*
*A Chronological Overview* . . . . . . . . . . . . . . . . . . *xvi*

**CHAPTER 1**   *The Defeat of the Sigmoid Colon* . . . . . . . . . . . *1*

**CHAPTER 2**   *Insurrection: 1950 Style* . . . . . . . . . . . . . . . . . *26*

**CHAPTER 3**   *An Early Education* . . . . . . . . . . . . . . . . . . . . *41*

**CHAPTER 4**   *Kid Newton and the "Mind-Body" Problems* . . . . . *49*

**CHAPTER 5**   *Eighty-Eight Yards of English Grammar* . . . . . . *59*

**CHAPTER 6**   *MOM–Or Methods of Management* . . . . . . . . . *80*

**CHAPTER 7**   *Summing Up: On Language* . . . . . . . . . . . . . . . *86*

*Addendum* . . . . . . . . . . . . . . . . . . . . . . . . . . . . *93*
*Memories of Thomas Gilbert* . . . . . . . . . . . . . . . *103*
*Photographs* . . . . . . . . . . . . . . . . . . . . . . . . . . *195*
*References* . . . . . . . . . . . . . . . . . . . . . . . . . . . . *201*

Preface to the Work

 Human Incompetence

# Thomas Gilbert
By Aubrey C. Daniels

I only met Tom on three occasions. Others told me of his irascible nature, but he was always quite friendly to me and complimentary about my work. You can't not like someone like that. When I met him I felt like I had known him for some time as his book, *Human Competence*, had shaped my thinking and writing since its publication.

Gilbert had a way of making the complex simple. His PIP is a classic in that regard. He took a concept like *competence* and created a definition that is immediately understandable and measurable. When many companies were struggling with trying to determine the mission of jobs, departments, and companies, we were teaching his ACORN test as a foolproof way to arrive at a clear, concise reason for the existence of a job, department, or company.

While there are many other ways that his work impacted our work, this book is more about Tom than about his work. Of course, it cannot give anywhere near a complete picture of the man he was. Describing him is much like the story of the blind men feeling the elephant. Each one "saw" a very different animal. As we have talked to many people who knew him as a friend and/or colleague, each describes a very different person. Since I only knew him casually, I have learned the most about who he was from his writing.

I know he had a great sense of humor. I am sure much of it was irreverent. Who doesn't enjoy his poems? I only wish that he had been able to complete this autobiography.

I love his stories, particularly those that he told about his time at the University of South Carolina and the University of Alabama. I am a native South Carolinian, so the way he relates those stories, I feel like I know the characters and developed vivid images of them. I have certainly known people from my youth like Barton Hogg who managed students collecting lead from the firing range at Fort Jackson. I know how he sounds and how he looks. I can see Bear Bryant explaining how to be a successful football coach. (You will have to read this book to learn how.)

Because Tom is concise and never one to "beat around the bush," he is quite quotable, and I have quoted him often. One of my favorite quotes from this book is, "Money is a beautifully honed instrument for recognizing and creating worthy performance. It is the principal tool for supplying incentives for competence and therefore deserves great respect. Any frivolous use of money weakens its power to promote human capital—the true wealth of nations." If he were alive today, he would be having a field day with the government's economic policies. He would no doubt have many clearheaded suggestions about how it could be done much better.

Although we miss the genius of Thomas Gilbert, his work lives on through his writings and through the work of his many devoted followers. We are honored that Marilyn Gilbert, his devoted wife, chose us to publish this book. It allows us to be a small part of his legacy.

*Aubrey C. Daniels, Ph.D.*
Founder, Aubrey Daniels International

Confessions of a Psychologist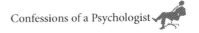

# The Publishing of this Work
By Darnell Lattal

*". . . how like the needle information is.
It always has a point and needs an eye."*

**Thomas Gilbert, Ph.D.**
*Poet and scientist*

This book is published as written. We understand that Dr. Gilbert did not intend this edition to be his final version. We know that there are unfinished thoughts and unclear reference to materials that should have come before or were intended to be elaborated on later; however, we made a decision not to ghostwrite for Tom.

Throughout this book, you will find mention of "more to come" at the end of some chapters. We kept that phrase, but sadly, he did not have time to write those additions to his story.

Tom was in the hospital when the typed and coffee-stained manuscript we were given was handed to his physician for review. It is unclear as to whether the doctor understood the genius of the man who had asked for his review. The doctor challenged a few of Tom's comments but without a great deal of visible excitement. If that discouraged Tom, we will never know. The manuscript, along with many other writings and works in progress, was put away, not surfacing until this last year when Marilyn, his wife, found the copy with the note to the physician and his edits, among other unpublished writings.

 Human Incompetence

Earlier, Tom sent a copy, very similar to this one but missing a few additional pages, to Alyce Dickinson to help her in a project she was doing to document the history of organizational behavior management. This seems to be, to the best of our knowledge, his last efforts. After the hospital stay in 1994, he kept these musings to himself, not yet asking Marilyn, to help him edit the work, as he had done with much of his writing throughout his life. His illness progressed rapidly and he died before he could do more.

Tom died in September 1995, and was in poor health for several years prior to his death—still much too young for all that he had yet to give.

Until she worked through a number of boxes long stored away, Marilyn was unaware of Tom's efforts to capture his life's story. He was however aware of her influence on that life. As his dedication says, she was his "everything." Marilyn contacted me at Aubrey Daniels International the year before, asking if we were interested in some of Tom's unpublished writings, including an expansion of his work found in the classic *Human Competence,* for our publishing house, Performance Management Publications. We jumped at the offer and have been going through his works with Marilyn's help, very carefully—they contain writings on instructional design, the next book beyond his masterpiece, *Human Competence,* and a strange and wonderful manuscript about knowing the animal within us all. He makes an oblique reference to that animal in a story you will find in this book. We hope to have additional works out in 2012 and 2013.

Confessions of a Psychologist

A few weeks before the International Association for Behavior Analysis meeting in May 2009, Marilyn called and said she had found another manuscript—his autobiography. Were we interested? Indeed we were. She brought it to me at a breakfast meeting at the hotel where we were staying and we discussed what she hoped we could do with the rather short work. We decided not to change what Tom wrote. We also kept his title. As you read the book, you will see why Tom titled it as he did.

Marilyn said that many had written about Tom's impact on instructional design and the value of his work from the perspective of engineering worthy performance, but few had written about their personal memories of the man. We decided to collect stories about this man from those who knew him personally. A short list was created at breakfast. That initial decision changed to adding a few others who did not know him but had been influenced by Tom's work and were well known to Marilyn in this regard. That list grew and continued into March of 2010. Additional papers, poems, family pictures, and other work relevant to this book continued to come in as well.

These personal vignettes by behavior analysts, engineers, psychologists, family, and friends do convey a picture of the man. There were others who wanted to contribute. Some who were asked were unable to do so due to other commitments, and a few, who had already written extensively about Tom elsewhere, declined. In seeking colleagues and friends, we missed some very significant people who were close to Tom. For a few, we could not find a way to connect and for others, we were unaware of the connection. For those of you we missed or whose comments are not included for a variety of reasons, I do apologize. We hope this original work, comments, and photographs help to bring this man to life for all who read this. None of this would have happened without Marilyn Gilbert's contributions and suggestions. She gave us editorial license to approach this as we wanted and she has been a great help in the production of the final work.

Brilliant, tough, blunt, aggressive, curious, bold in his creative reach, innovative and entrepreneurial, wise about the human condition, impatient with platitudes—very few gilding of any lilies for Tom—a robust sense of humor, unwilling to edit his comments to fit convention, willing to laugh out loud and to let you know when he disagreed—out loud—those are some of the attributes I now associate with him. Cook, loving father, delighted granddad, and reader of mythology, great literature, and poetry are a few more. Tom was not visibly a sentimentalist, but there are moments when his sentimentality comes through—his appreciation for family and for emotional connections, demonstrated publicly in a description of his bragging rights about the beauty of his wife, at his visible reactions at a wedding of one of his children, and at the birth of his granddaughter. He was a lover of "rhyming" poetry. He wrote poetry. He expressed deep affection for his children—and who they were becoming. He was delighted to be a grandfather. He wondered about how to raise them; asking advice and editing advice. He knew the limits of procedurally correct but uninformed discipline as a teaching tool for children. He provided adventures (camping out in the backyard, cooking large meals) and great humor along their journey. He wanted good things to happen to them. He spent his leisure time with them and with Marilyn. He established robust friendships and sustained them throughout his life. He was busy at his work until he was too weak to work. He had much to say and too little time to say it.

Ray Fowler, former president of the American Psychological Association (APA) and then executive director for many years, said that Tom Gilbert was one of the top three most brilliant American psychologists of the twentieth century. His genius is worth celebrating. He changed the way we look at and strive for excellence in human performance.

I read about Tom in his own words and as those who knew him wrote about him. They will tell you about Tom—unvarnished. For me, who knew him very slightly, this has been a journey of discovery. Thank you, Marilyn, for this incredible privilege. We hope we have done him justice.

**Darnell Lattal, Ph.D.**
CEO & President
Aubrey Daniels International

## Foreword by Marilyn Gilbert

*"Here with a Loaf of Bread beneath the Bough,*
*A Flask of Wine, a Book of Verse - and Thou*
*Beside me singing in the Wilderness -*
*And Wilderness is Paradise now."*

## The Rubiyaiyat, by Omar Khayyam

Tom left me many unfinished books about his work, some of them still in his careful cursive and on lined, yellow paper. After a while, he probably could write the first drafts in his head. I've gathered those book starts and some typed, but unpublished, writings and sorted them into two books, which Aubrey Daniels International will soon publish. One is on instructional design, and the other continues performance engineering. All the writing is original Tom, minus his typos and repetitions.

But Tom really did intend to finish this brief autobiography, and he dedicated it to me. I loved that. I thought it would make a great trade book, perhaps inspiring to young adults. That possibility, however, would have required editing. And when I presented the manuscript to Darnell Lattal and Aubrey Daniels, we three agreed that an autobiography should remain intact and exactly as written. So, that's how it is.

There never was any halfway measure about Tom himself or how I felt about him. No one (except for my father) had ever read to me before. I listened to Tom read the lessons of Soren Kierkergaard (on an airplane), *The Rubiyaiyat* of Omar Khayyam (with wine), and some classics I never would have read on my own—Milton's *Paradise Lost*, for

 Human Incompetence

example. No one has ever made me laugh so often. No one has ever been so generous to me. And no one has ever taught me more—including how to enjoy *Archy and Mehitabel*; how to appreciate Dali and Mirot; how to hear Hank Williams and Johnny Cash; how to watch Joe Namath; and how to tolerate the New York Yankees. Sometimes, when I'm stuck on a problem at work, I still cry out, like *Mad Men's* Betty Draper in labor with her third child, "Where is Tom when I need him?"

All who loved him miss Tom—his children, his stepchildren, the grandchildren who were lucky enough to know him, and of course his friends. Our first grandchild, Georgia, misses Tom in a special way. He taught her to say "her colors" and to recite the alphabet. And he made sure that her drawings were on permanent exhibit at the local pizza parlor in Clinton, New Jersey. She was almost six when he died, and I can only imagine how proud he would be if he could see her as she is today, almost twenty-one.

Georgia loved Tom unconditionally. One sunny afternoon, Tom fell in the front yard, and Georgia and I couldn't get him up and on his feet. I ran into the house to call 911, and Georgia ran with me. When I returned outside to wait for the ambulance, Georgia was holding a large rain umbrella over Tom. "I don't want him to get sunburned," she said.

But even with Georgia, Tom remained true to himself. As he waited for her after nursery school one afternoon, he darted into the children's toilet at the rear of the classroom; very soon, curls of grey smoke floated out from the opening under the little stall. Miss Jenny said nothing. The next day, she implored me to control him, please—as if anyone except Superwoman could do that. But later, when she heard Tom was terminally ill, Miss Jenny stopped me in town to say she wanted to care for Georgia after school, at no charge. No wonder Tom loved New Jersey and our neighbors there.

When Georgia was in high school here in Washington State, she read *Slaughterhouse Five,* Kurt Vonnegut's story of his own horrific experience as a prisoner of war in World War II and witness to the destruction of the city of Dresden. If you haven't read the book, Vonnegut's main character, Billy Pilgrim, went time traveling and visited the land of the Tralfamadorians whenever events in his own life became overwhelming. Here is what Georgia wrote:

"The Tralfamadorians believe that when a person dies, he only appears dead but he is still living in other moments in history, so he's not really dead. They can look back on events of the past as if they were happening at that very moment and then act in them.

"In this passage, the main character, Billy Pilgrim, explains the two separate views of humans versus the Tralfarmadorians, saying that the way humans deal with death is pointless because they [the dead] are still there in other moments of history. If one focuses on the death of another, he or she will also forget all the influential moments in that person's history and how these moments have affected others around them.

"When I was about six-years-old, my grandfather, who was my main father figure up until that point, died of lung cancer. After he was gone, I focused mainly on his death, and

 Human Incompetence

not on the moments we shared that changed my life and helped form who I am today. It made me a different, less stable person than I could have been; and now, some 11 years after his death, I wish I had known about this view of dealing with death since I probably could have saved myself a lot of hardship later on. Even though it's hard to deal with the death of a loved one in that manner, it is better off for the person in the long run.

"The book talks about many people who have died who are influential to everybody and to nobody, and in response to their deaths, the author says, 'so it goes.' This obvious display of distance from person to person will help the author in the long run to overcome the fear of death that many people face. He probably feels that when he dies, his death should be viewed in the same manner, and that it doesn't really matter because he's not really dead."

So it goes.

I am thrilled by the number of people who have written their memories of Tom, and thank you all! I hope I didn't miss others who might have wanted to contribute. And I have a special thank you—to Darnell Lattal, Aubrey Daniels, and Laura-Lee Glass—for bringing Tom back in this way to honor him and his work.

*– Marilyn Gilbert*

Unfortunately Tom wrote little about his family in this unfinished autobiography, but we found a note that captures the busy man described above. This note to Marilyn, his "everything," provides a glimpse into the private Tom. We thought you might enjoy reading this message, in particular his handwritten note at the bottom. Marilyn, we might conclude, was a very patient woman.

— *Darnell Lattal*

**[SIC] Estamos: we are ready**

**3-29-95**

*Dearest Girl:*

*Here is a draft of a list of improvements I shall aim for over the next few months. A progress plotter accompanies them—without which they are northern. Care to comment or add to them? You might want to make your own list; I gather that is the way it's supposed to be done.*

- *By the end of April—if not sooner—I will discuss nothing but contemporary events. No more dragging up old, tired stories.*

- *I will make considerable progress in the Job Aid business with you and Byron. This requires a separate progress plotter, which I shall soon make.*

- *My office will be spick-and-span by May 1st.*

- *By June, I shall complete the Children's Clock and Progress Plotter, with or without Mark's help.*

- *Sometime in April I shall be taking regular initiative in planning pleasant activities we can do together.*

*Love,*

*Your improving boy*

*I need some suggestions about the intro to measurement! Help, Help!*

 Human Incompetence

A Chronological Overview:
Gilbert's Biographical Milestones

As I mentioned, this autobiography was unfinished and incomplete, leaving many questions about the exact time frames and events of his life. We could not find any biographies of Tom's life that outlined in detail the summary of that life—the where, when, and what of various times and life experiences. Marilyn Gilbert has written an overview that will help you as you read the autobiography. She has written it as she experienced parts of it and as she heard from Tom about aspects of his life, especially the early years.

It is another perspective to add to what you will discover about this man as you read his work and the words of others who knew him.

*Darnell Lattal*

# Tom Gilbert's Life as I Know It

## Marilyn Gilbert

Thomas Franklyn Roby Gilbert was born in Durham, North Carolina, on January 3, 1927, to Lillian Barton Gilbert and Franklin Gilbert. It was a challenged birth, since Lillian, eight months pregnant, had been hit by a city bus and was hospitalized for the remaining weeks of her pregnancy. She was a stay-at-home mom at first, as Frank supported the family in a succession of jobs in circulation departments of small-town newspapers across the state of North Carolina. Tom's brother, named James after Lillian's father, was born when Tom was almost five. And his birth at home in Elizabeth City, North Carolina, was an event Tom never forgot. Perhaps the Gilberts had planned a trio of Tom, Dick, and Harry, since James was always called Dick; but Harry never happened.

*Tom's father, Frank Gilbert, as a young man*

As the Great Depression advanced, and Frank faced steady unemployment, the family returned to Lillian's ancestral home in Columbia, South Carolina. (Frank was born in Tennessee.) Lillian's mother, Martha, managed a boarding house, and Lillian's father, "Uncle Jimmy," patrolled the streets of Columbia as a trusted

Human Incompetence

member of the Columbia Police Department. But good jobs were scarce in South Carolina as well. Of Lillian's five siblings and Frank's twelve, all faced poverty except for Lillian's older sister Eve, who was rich by the yardstick of those times. (Eve had managed to meet and marry a succession of wealthy men.)

Now Eve lived in a large house in Long Beach, California, near Los Angeles. Frank decided he would take the family west and they could stay with Eve while he searched for work. He also persuaded his sister Jean to join them, since her husband owned a small truck that could accommodate both families. All seven of them set out on the great trek in pursuit of a better life, like many Americans at that time—most of them Southerners. On the road all day, they stopped only for their meals or to repair the truck, a faulty wheel or gasket. At night, they rolled out the blankets so the grown-ups could sleep under the stars as the children slept safely in the truck. Tom remembered just bits and pieces of that trip, or perhaps he remembered *Grapes of Wrath*. But at the end of the long journey, Eve welcomed the travelers to her home as planned.

The search for work proved fruitless both for Frank and Jean's husband, and Tom's memories of California were mostly bleak, unpleasant ones. He recalled his panic as his father carried him out of the house during the Long Beach Earthquake in March of 1934. He felt ashamed as he watched his mother beg on the streets of Los Angeles. And he felt angry when his schoolmates taunted him for accepting the government-provided milk (it was called "relief" then).

After about a year, I'm not certain of the exact date, Frank and Lillian brought their family back to Columbia. But I do know that Tom started school in California, and he was back in Columbia at least by the second grade, since that was where he met his first wife, Betty (Elizabeth) Battle. Tom was happy to reunite with his relatives in Columbia, and particularly Martha and Uncle Jimmy and his Aunt Myrtle. He saw his paternal grandparents occasionally, but he remembers "Mr. Gilbert" as being too stern. (I suppose it was a Southern custom for elderly Southern wives to refer to their husbands by their surnames, but I was always amused by these references.) "Mr. Gilbert's" first wife, mother of the first eleven Gilbert children, had died before Tom was born. But his second wife, the mother of the twelfth Gilbert, was straight from Scotland, and Tom remembered her and her brogue very fondly.

With the economy somewhat improved, Frank found work in Columbia, probably in the Circulation Department of *The State*, Columbia's primary newspaper. Tom said Frank had a great radio voice, and on Sundays he sometimes read the funnies for a local radio station, in the image of Mayor Fiorello LaGuardia reading to the children of New York during a newspaper strike. I never met Frank, but I supposed he was very intelligent, though uneducated. Relatives said he was charming, complex, and very inventive; he invented Welcome Wagon, in fact. Also a passionate atheist, Frank "wrote" a large book of empty black pages titled *The After Life*.

## Human Incompetence

*Home in Dentsville, South Carolina*

Lillian, too, found work in Columbia after taking a typing course her sister Eve had won but had rejected for herself. At first, Lillian worked as a typist—she was very skilled. But she was also smart, ambitious, frugal, and very practical. Her passion became real estate, particularly ocean property, which she always said, "they aren't making any more of." She started by buying a little house in Dentsville, outside of Columbia. She also owned one of the first motels on the Interstate Highway. (In the North, we called them "Tourist Courts.") Mornings after the guests left, Lillian would put Tom to work changing the sheets and cleaning up the rooms. I gathered that he didn't perform these duties with due diligence, however. By the time World War II began, Lillian had sold the little motel and now owned a cleaning store at Fort Jackson. Tom worked there too, even sewing the Eisenhower jackets. I'll confess that he did all of our ironing and sewing too, including sewing buttons on my blouses, because he was

*Tom and his mother Lillian looking at a globe when he was about nine years old*

Confessions of a Psychologist

better at it than I was and he minded doing it less. It was no surprise that Lillian and Frank were not a happy team: they were much too different. I also surmise that Frank had become discouraged and drank too much. Tom was seventeen when they divorced. Afterwards, Frank lived very quietly with a partner, Marian, until he died. Lillian married a golf professional from New Jersey, Mike Serino, who had survived the Battle of the Bulge. At first, they lived the lifestyle of professional golfers, and Mike gave golf lessons. He taught Tom and Dick to play, and Dick became an accomplished golfer. But Mike's most famous student was Dwight David Eisenhower, when the general was stationed at Fort Jackson. As Mike grew older, he left golf, and he and Lillian retired on her income. After Lillian became ill, Mike secretly took up with a much younger woman. And when Lillian's bank notes came due, this woman forged Lillian's signature. The two married a month after Lillian's death and lived in Lillian's house in Myrtle Beach. We never saw Mike again.

I should say here that what I know about Tom's childhood is anecdotal, just what he told me in bits of conversation and also some hearsay from his Aunt Myrtle. So, now that I'm trying to arrange the happenings in some sort of order, I may be off the mark. But sometime while he was still in elementary school, and before the invention of antibiotics, Tom contracted the serious bone disease called osteomyelitis. His recovery was slow, and he had recurring bouts of these infections throughout his life. It was the same disease that plagued the life and baseball career of Mickey Mantle, Tom's idol. Tom loved baseball and was a huge Yankees fan.

Human Incompetence

Because of his problematic health, Tom was 4F during World War II. So, after graduating from Columbia High School, he entered the University of South Carolina, where, from what he told me, he was an indifferent student. If he liked a subject, he got an A+. If he wasn't interested, his grade was C. Betty Battle, his friend from second grade, was also a student at the university. And after they graduated, Tom and Betty were married; that was in 1948, when both were twenty-one. They went off to Knoxville, Tennessee, where Tom eventually earned a Ph.D. in psychology at the University of Tennessee. I won't try to describe those years, since Tom has written about them himself and, besides, I wasn't there.

*High school picture of Tom*

I first met Tom at the historic Bell Laboratories, in New Providence, New Jersey. It was August of 1958, and my husband—Charlie Ferster—and I had driven from our home in Indianapolis to visit our parents for the week. On Saturday, we planned an outing in New York City with our old Cambridge friends, Adair and Herb Jenkins, who had recently moved to New Jersey. As early supporters of Fred Skinner and his research, scientists at the "Labs" hired Herb to set up a pigeon laboratory and do research there. Charlie was eager to see Herb's setup, of course, so we stopped at the Labs first. But it was Tom Gilbert who greeted us at the door.

Confessions of a Psychologist

*Tom graduating from the University of Tennessee, 1949*

In September, Tom would go to Harvard, on a post-doctoral fellowship, to study and develop instructional materials for the teaching machine. That summer however, Fred Skinner had arranged for Tom to work at Bell Labs to solve a problem with tiny electronic parts called resistors. Resistors came in different sizes; but because they were so small, size numbers would have been too hard to read. So instead of numbers, resistors wore color bands—different colors for different sizes. Bell Labs had wanted Tom to write instruction that would teach this color code, and Tom delivered an unusual one-page lesson: a set of mediators he had devised, linking the size number to its color. For example, 1 brown penny linked 1 and brown, 2 hearts linked 2 and red, and so on. It was a wonderfully inventive system, requiring just a few minutes to learn.

But, at that time, I knew nothing about Tom or his work. When Herb appeared and whisked Charlie off on the tour, Tom stayed to entertain me. We talked mostly about the South. I don't know why, except that Tom had probably introduced himself as a southerner. Besides, Charlie and I had spent two years at Yerkes Institute of Primate Biology when it was in Orange Park, Florida, and much of what I had observed there still puzzled me. I remember asking Tom why the band played "The Battle Hymn of the Republic"—a Yankee favorite!—at a football game between two Southern universities. Tom smiled, saying that Southerners typically had special lyrics for Yankee music. When I asked what these lyrics were, he demonstrated: "We'll hang Abe Lincoln on a sour apple tree; we'll hang Abe Lincoln on a sour apple tree..." I hardly expected that. Tom was a proud Southerner who was also a progressive. We had both cast our first vote for candidate Norman Thomas, he in South Carolina and I in New Jersey. And later, he told me about unpleasant treatment in Boston from Yankees who thought all Southerners were racists.

That night, the Jenkins's babysitter canceled on short notice. So, Adair invited all of us for a backyard barbeque. It was a beautiful, starry evening, and in the fashion of those times, we drank excessive martinis. I have never forgotten Tom's farewell salute to Adair and me: "You beautiful queens, you beautiful queens!" We laughed at him, but we were both charmed as well.

The next day, Charlie and I returned to Indianapolis, where Charlie was a member of the new Behavior Research Institute at the University of Indiana Medical School. I began to write children's books on math and science for a small publisher, and I taught English Composition at IU's satellite in Indianapolis. Charlie and I had already started the *Journal of the Experimental Analysis of Behavior (JEAB)*. He was the first editor; I, the first associate editor. I had typed the first two issues on a proportionately spaced IBM typewriter, producing the justified margins of hot type by counting the characters and retyping to adjust the space.

Meanwhile, Tom went to Harvard as planned, moving his family from Athens, Georgia, to Lynn, Massachusetts. In Memorial Hall, he shared a workroom with Susan Markle, who was also exploring instruction for the teaching machine. Tom named these products *programmed instruction*, probably because their structure seemed so rigid. Later, Sue wrote that the proper word was *programed*, because *programmed* was grammatically incorrect. Perhaps, but the early name has stuck.

Although Tom attended Fred Skinner's Friday afternoon pigeon meetings and he was sometimes outspoken there (I'm told), his primary interest was in applications of behavior analysis—*operant conditioning*, it was called then. He spent a lot of time with Og Lindsley, who was doing research with inmates of a mental hospital outside Cambridge, and Tom and Og became good friends. He also developed a firm friendship with a young Harvard instructor, Charlie Slack, who was working with teenage male, delinquents in Boston.

During this time at Harvard, Tom wrote the now-famous "Behavior Research and Development Proposal," which was a hugely extensive plan to set up laboratories for scientific behavioral research and the development of applications of behavior analysis. I also have a fat binder with all his writings then. Charlie Slack became partners with Tom and left Cambridge. Tom didn't return to his teaching position at the University of Georgia either. Both of them moved their families to Tuscaloosa, Alabama, where they hoped to build the proposed institute once it was approved and funded by the National Institutes of Health (NIH). Tom had chosen Alabama because he was able to garner support from high political figures there, through good friends like John McKee, who worked as a psychologist at Draper Prison, Alabama, and also for the state. Another enthusiastic and influential Alabama supporter was Paul Siegal, chair of the Psychology Department at the University of Alabama. Paul hired both Tom and Charlie Slack to teach psychology so they could support their families while waiting for the nod from NIH. And Paul's later reminiscences of his experiences in managing those two exceptionally unconventional characters are hilarious.

The next I heard about Tom was when his phone call to Charlie interrupted a dinner party at our house. Tom offered Charlie the position of president of the Institute, a position Fred Keller and others had already refused. But Charlie was intrigued and considered the opportunity seriously. I was not happy about a potential move to Tuscaloosa. But if Tom's dream had materialized, and Charlie had accepted Tom's invitation, I surely would have gone with him as dutiful faculty wives did. But the dream did not materialize for anyone, unfortunately, and nothing like this program has ever been proposed since. I also suppose there never was a time like the late fifties and early sixties, when people were so willing to invest in education. Looking back at it now though, I can underttand why nothing happened. The *Great Educational Revolution* was more words than deeds.

After losing his dream of the Institute, Tom hunkered down to further study the design of instruction. In an often-quoted speech at a convention of psychologists, Tom advised, "Throw away your teaching machines. If you have an urge to get one, buy a toaster instead."

But his disillusion was more with the lack of design in the instruction itself than with its delivery by a machine. The lessons were ineffective, just linear streams of content—little pieces of knowledge in small, defined physical spaces that Tom himself had named *frames*. The only evidence of science was in the immediate feedback that the correct answers afforded. In contrast, Tom believed that the amount of physical space for teaching a concept should depend on what the instruction needed. A novel idea! The system he was developing had many other scientific features besides immediate feedback, but it wasn't yet nearly as complete as he would make it later. His model was the set of scientific principles used in training pigeons and rats in the environment of the Skinner box. He called his system *mathetics*, after the Greek word for *learning*.

Although I heard some bits of information about Tom's work, the next time I actually saw Tom was after Charlie Ferster invited him to talk about mathetics at the Institute's monthly symposium. Charlie was in charge of acquiring the speakers that year. And since Charlie was the lone behaviorist at the Medical School, all the speakers he selected were behaviorists. Our routine was to throw a party after the lecture and to provide sleeping quarters for the guests. Bea Barrett met Og Lindsley at our party for Og when he was guest speaker. (Bea left Indianapolis soon after, became a behavior analyst in Boston, and never looked back.)

When Tom arrived for his talk, it was the summer of 1960, hot and humid. Whoever wrote the announcement of Tom's lecture, which was posted around the Institute, had typed "mathematics" instead of "mathetics." (This is actually a common typo, and even a common misreading when it is typed correctly.) So, whether it was the result of the weather or the typo, attendance for Tom's lecture was embarrassingly low. I was there because I wanted to learn more about designing instruction. As the textbook for my classes that spring, I had selected the new *Programmed English 2600* and found it ineffective. But at our house the next day, Tom seemed undaunted by the low turnout at his symposium, and he talked excitedly about mathetics. Charlie and I became excited about it, too. And when I declared I would use mathetics to write a book on writing, Tom dubbed me Matheticist Number I. After he left, I did try to write a book on writing. Like most teachers, I knew the content but very little about the principles of learning. That desire to develop a "better book" on writing though, became my lifetime passion.

Some time later, Tom decided to start a journal for applications of behavior analysis, and he invited Charlie and me to Tuscaloosa to query us about our experience with *JEAB*. Not long after that trip, we received the first issue of the *Journal of Mathetics*. The model appeared to be Tom's theory on instruction plus invited applications of other behavior analysts. (I have to wonder now if the *Journal of Applied Behavior Analysis—JABA—*has ever considered Tom's journal their forerunner. In the January issue, the invited article was Charlie's "The Control of Eating," describing a weight-reduction program he had undertaken with two other researchers. Charlie's words were the only words in the first issue that Tom did not write—not that he ever admitted

*Tom and friend Carroll, 1964*

this in the journal. The lead article was his "Mathetics: The Technology of Education." Tom also wrote the other two articles, but attributed authorship to others. He named S.K. Dunn as author of "The Progress Plotter as a Reinforcement Device." And he named L. Jungberg Fox—who was the Swedish uncle of Don Cook's wife, Dawn—the author of "Effecting the Use of Efficient Study Habits." He also named me as the Associate Editor, even though I had done no editing for him.

Tom's lead article on mathetics was brilliant, although flawed in some ways. Yet from there, I can trace the development of Tom's work before *Human Competence* and after. I am always surprised that most practitioners of Human Performance Technology do not recognize that Tom's major accomplishment was his contributions to the design of instruction. I suppose he made his point (too well) that other components besides lack of training contribute to the quality of performance.

Tom and Charlie Slack continued to attend conferences. At one of them, they met a distinguished publisher of medical books, Bernie Springer, who wanted a textbook that would address a problem in the education of nurses. As a group, nurses were not good at math, yet they were often required

*An early picture of Tom taken in 1964 by Phil Niblock*

to prepare medications in metric units. The book would teach them how to convert our English system of pounds and ounces to their equivalent metric units. Remembering my desire to apply mathetics in a book on writing, Tom called to ask me if I wanted to write *Arithmetics for Nurses* instead. I was delighted. As an undergraduate, my double major was Latin and mathematics. I had no unusual ways to teach math, and I didn't know a lot about mathetics at that time either. So I modeled my book on a program Tom had developed for teaching long division—except that I didn't start from the end of the chain as Tom had done. That sequence never made sense to me. After my book was published, it soon became a model for other books on this same topic, although none had existed before. The authors of these books mostly copied what I had done, including my errors. It was a great learning experience.

During this low period for Tom and Charlie Slack, they adjusted their lives significantly. Charlie and his family left Tuscaloosa for New Jersey, their hometown. Tom stayed in Tuscaloosa and started a small consulting company to produce mathetical products. At first, he hired and taught some smart psychology students and graduates to work for him, including Betty Pennington and Betty Hatch. As he continued, he was able to lure an instructor of psychology, Dempsey Pennington, away from the University of Alabama. He also petitioned Robert Reynolds, a graduate student he knew from Georgia, to manage the whole operation.

Meanwhile, he and Charlie Slack continued to pound their message, seeking seed money to develop the larger company they envisioned. Somehow, somewhere, they met Carl Sontheimer, a Connecticut engineer who had achieved financial success with a small electronics company he had developed. Carl now wanted to use some of this money to improve education, and Tom's ideas strongly attracted him. Carl set up a mathetics company in Connecticut which Tom named TOR, and both Charlie Slack and Tom moved to New York without their families and worked in Connecticut.

I don't know how long Tom lasted at TOR, since I was not a part of this. I never even learned exactly what happened and why he left. I do know that TOR had a life afterwards, without Tom, as a home study school in Chicago. I also know that Carl sued Tom for ownership of the April issue of the *Journal of Mathetics*. Tom won the suit however, and paid for the publication of the April issue himself. It was shorter than January's issue, with only two articles. The lead article was "Mathetics: II. The Design of Teaching Lessons." Israel Goldiamond wrote the second article about his work on stuttering called "The Maintenance of Ongoing Fluent Verbal Behavior." There never was a third issue, but the first two issues still have a full life, at least for Kent Johnson and the staff at Morningside Academy in Seattle.

# Human Incompetence

The investor Carl Sontheimer quickly found the grand fortune he had desired, but not in education. Morbidly obese and a talented cook himself, Carl Sontheimer reinvented the food chopper by electrifying it and marketing it as *The Cuizinart*. It became wildly popular and probably made Carl a billionaire several times over. His company of that same name soon broadened to include all sorts of kitchen tools, even coffeemakers. But a Cuizinart never graced the kitchen of Marilyn and Tom Gilbert. Tom said only, "I wouldn't buy oxygen from him, even if he was the only one selling it."

After TOR, Charlie Slack moved on, eventually settling in Australia. Tom went to Atlanta with his family, and he consulted full time. We didn't hear from Tom at all until Charlie and I moved to Silver Spring, Maryland, when he made a surprise visit. He had a broken leg and was on crutches, and he appeared depressed and without his usual exuberance.

In Maryland, as problems in my marriage became more serious, Charlie and I separated. I worked on freelance writing assignments. I also succeeded in writing the first of my several books on writing. This one was called *Programmed English Composition*, and Appleton Century Crofts published it in 1965 as a textbook for English Composition classes. Another Charlie, Charlie Walther, who was Fred Skinner's longtime editor and champion, was my editor, too. But my book was not the splash I had wanted, although it did have some unusual features. One of these was to show edited text with the copyediting marks. This feature is now commonplace in books on writing.

Tom's marriage was also broken. Although there had been separations earlier, the breakup this time was permanent. He went to New York, and that's where we got together. At first we worked on separate consulting assignments. But after a few months, Tom suffered unexpected health problems and was diagnosed with cancer. (That diagnosis later proved false.) Besides, we both still had our "problems of living," as he always put it, to work out. He wanted us to go to Mexico City—to essentially drop out. He was resolute. Since I didn't have much to lose either, I was easily persuaded to join him. So, on April 1st of 1965, that's what we did: we dropped out.

We drove in Tom's car to Quernavacca, Mexico, stopping along the way to visit museums and swim in the Gulf of Mexico. We toured America's Southland and then Northern Mexico. In Quernavacca, we met an artist from Massachusetts, Robby Goodale, and his lovely wife, Irene. Tom and Robby were incredibly alike, and in one week they became best friends for life. We rented an apartment in Mexico City, where we stayed for five months—reading, trying to learn Spanish, watching bullfights, working on some projects we received from New York—until we became homesick. Still not ready for New York, we moved on to New Orleans and stayed there for another three months, surviving Hurricane Betsy, but running out of money. Fortunately, Tom, with the help of his good friend Byron Menides, cashed some TOR stock that Byron had long ago persuaded him to buy. On Thanksgiving Day, we boarded a train for New York. We had somehow recovered both ourselves and our interest in instruction. Tom and I were married in New York's City Hall, on May 5, 1967. Although we hadn't thought about

# Human Incompetence

*Tom in 1965: busy at work; a happy and productive period in Tuscaloosa, Alabama*

it before, our wedding date was Cinco de Maio, the day we had first arrived in Mexico, finding everything closed for the holiday.

In New York, we earned our living at first by freelancing. But slowly we renewed old connections and made new ones. Tom set up a little company with still another Charlie, Charlie Jacobs, a lawyer. (Wouldn't you know it—my father was a Charlie too!) Tom called the new company "Praxeonomy," an ugly name. But after an earlier colleague had succeeded in copyrighting the term "mathetics," Tom wanted to use a name so ugly that no one would snatch it from him. Praxeonomy's offering was a giant step forward in Tom's development of instruction, though, since now it was not just behavior analysis: it was performance analysis—starting with accomplishments and developing them by using principles of behavior analysis. But, like most things new, our workshops were a very hard sell, and our announcements pulled in very few paying takers.

Not long after Praxeonomy folded, however, Praxis was born. At the start, there were three principals: Tom was President, Geary Rummler was Project Manager, and Irving Goldberg was CFO; they were backed by an accountant, Marty Mensche. Praxis probably spent too much money at the start, in the flush of new money and renewed confidence in the product.

Fortunately, at some conference, Lee Brown, a Chicago publisher of "books in boxes," found Tom Gilbert, or Tom Gilbert found Lee Brown. I don't remember which is correct. But soon after their fortuitous meeting, Lee moved his family to New York, as well as his business, and he was prepared to spend his life funding Tom's products. Tom responded by doing some of his best work so far, starting with his beginning reading and writing program. Tom succeeded in teaching four-year-old children on Manhattan's West Side, as well as four-year-old children in Harlem, to break the reading code, with what was probably a semester's worth of instruction. That design would have been perfect for the computer. The plan was to continue through eighth-grade reading and writing.

*Marilyn Gilbert, busy at her editing tasks*

Another Lee Brown project for Tom was to develop a social sciences curriculum, and Tom had begun the research and the beginnings of the design. Lee planned for Tom to develop a complete school curriculum, K through 12. But Lee Brown was interrupted by life—rather, the loss of it. One Christmas vacation in Mexico, one of his three daughters accidentally drowned. And on a subsequent Christmas vacation at a ski resort in Colorado, Lee took a nap after dinner and never woke up. His death was a terrible loss to Tom and me, but especially to Tom, because they had great respect for one another and shared a wonderful professional relationship. Had Lee Brown lived, I am certain that Tom's work would have taken a very different path. I sometimes wonder where that path might have led us.

Lee Brown's company of two principals, Lee and his editor Dee (I forgot her last name) quickly dissolved and Tom's projects with it. Although Praxis tried to pedal Tom's reading program to other publishers, no takers stepped forward. Much later, Tom offered the reading program to traditional publishers again. But these publishers wanted a finished program from K to 8, without making any investment in development. Even then, few conventional publishers would take chances on unconventional products. Lee Brown was special. Although Tom's design for teaching the social sciences wasn't well enough developed to present to publishers, Tom did write an article about the process of choosing the content of a subject matter. He called this article "Saying What a Subject Matter Is." It has been a tremendous thought provoker for me and a strong influence on my work. I thought it should be rewritten, though, to include what Tom left out, probably because he didn't bolster his narrative with the examples that actually existed in his own work. These writings, with examples, will be in a book that I hope to deliver to Aubrey Daniels International in the next months.

Praxis continued with other clients besides Lee Brown, and Tom designed ingenious solutions to their problems. Many of these clients were government agencies, including Social Security, the Forest Service, and even the IRS. Banks were

Confessions of a Psychologist

also clients; so was Xerox, so was the Marriot Hotel, and so were a few drug companies. My favorite of Tom's projects then, although I didn't work on it, was with a small company in Philadelphia. The company wanted a course teaching elementary school children how to play the clarinet, the most popular instrument in school bands across America. This project tickled me from the outset because Tom was almost deaf in one ear, didn't hear too well in the other, and was totally tone deaf. Country music was the only music he enjoyed, but because of the lyrics, not the tunes. Never mind that. Tom's design for achieving the performance required was so radical, yet so on target, that he delighted top musicians at the Curtis Institute in Philadelphia, where he piloted the program. His invention for better breath control, as an example, was a physical feedback aide he called the Barrogauge. With this device strapped to their chests, students could easily tell whether they were breathing correctly or not. It was a different kind of training wheels. Later, when Tom was in the hospital after surgery, a nurse gave him a Barrogauge to help him breath properly and prevent pneumonia. Tom learned that his former client had secretly sold his design. Although this client had escaped jail after stealing the Barrogauge, he continued his criminal behavior and was caught and was now in prison for some years.

In the early years, I worked with Praxis on many projects, and I edited all of Tom's writings. Whenever someone at Praxis bailed on a project, I finished it. I also helped Tom develop and deliver the "Performance Analysis and Instructional Design Workshops." I also worked on projects independent of Praxis. One of the major early investors in better instruction was AT&T where I worked onsite two days a week as a kind of aide for the managers, either helping them to write their letters or instructional materials or editing after their writing was done. (How I would like a job like that now, here in Seattle or anywhere!) Praxis had many contracts with AT&T too, as did many of the early workers in the field.

In 1971 when Praxis lost its lease in a building on 13th Street, there was buzz about moving to New Jersey. Geary already lived there, and Irving was no longer with Praxis. I didn't like that idea at first, because I was born and raised in New Jersey, and I didn't want to return. But I had also soured some on New York. We had a small place in the Berkshires, but we were never able to spend longer than weekends there. Perhaps a different, less crowded part of New Jersey might be nice. So Tom took me back to New Jersey one day, and we found a house we could afford on 7.5 acres outside of Morristown. We lived there until our children were nearly ready for college. Tom loved New Jersey.

When Tom was recuperating from back surgery, I convinced him to write the book that became *Human Competence*, which McGraw-Hill published in 1978. I also convinced him to write the book with me that became *Thinking Metric*, published by John Wiley & Sons. I gave up my right of first author because it was Tom's design that made the book.

For a while in New Jersey, I worked full time at Praxis, as ghostwriter for Tom's "Levels and Structure of Performance Analysis" and much of the monograph "Knowledge Maps." And for a contract between Praxis and AT&T, I developed the "Course Developer's Workshop," which was an enormous affair with sample materials that included instruction on flowcharting. But after a while, I maintained contracts separate from Praxis, and I'm glad I did that, because it supplemented our income when Tom was ill and after Praxis was sold.

*Tom and Marilyn's home in Morristown, NJ*

The sale to Kepner Trigo in Princeton, happened in 1980 or so, with Tom an unwilling partner to it. Tom suffered a month of depression on our couch before we started over with the Performance Engineering Group. As a group of one, I still maintain the title. Tom did all the marketing, which was a new responsibility. Fortunately, some Praxis clients who had worked closely with Tom wanted to continue. Mike Barber was one of them and a frequent caller. Tom established the procedures in Mike's hardware store, and Mike experimented with Tom's schemes for incenting workers. Another special client for Tom was Shell Oil, where his assignment was to improve the job of design engineers. Pennsylvania Electric in Allentown called Tom to solve many of their problems.

Human Incompetence

Tom at home in Morristown, NJ, 1975

Tom also worked with a tobacco company and a cancer prevention agency. He was amused that no one at the tobacco company smoked, although cigarettes and ashtrays were available everywhere; at the cancer prevention society, there wasn't an ashtray anywhere, so all the employees took their smokes outside: another of life's ironies. We both worked on a huge project for AT&T, developing the new operator training, and we were invited to the New Year's Eve party of 1984 when the Bell system companies broke away from Mother AT&T. We also worked for the New England Bell Company on Yellow Pages sales, as well as for Towmotor Corporation on sale of lift trucks. I continued to write books—on writing for John Wiley & Sons and on SAT analogies for a small company in Piscataway, New Jersey.

Tom won the highest awards in the field and was often a guest speaker for groups interested in improving performance.

In 1983, we were invited to South Africa for six weeks, to

deliver workshops for the Productivity Institute in Pretoria, Johannesburg, Durban, and Cape Town. Apartheid had not yet been defeated, and violence was frequent. But South Africa was a beautiful country and a most fascinating experience for us. Before we left, we were treated to a long weekend at Kruger Park, for a photographic safari I will never forget. Shortly after this trip, we sold our house and moved to the home we shared in Hunterdon County, New Jersey, which we both loved. In 1988, our granddaughter Georgia was born and lived with us there, becoming the star in Tom's life. He loved teaching her, and she loved learning whatever he taught her.

Our last business trip together was to Sydney, Australia, another marvelous experience, where Tom gave workshops. However, Tom's work experiences afterwards were limited and not always stellar. In August of 1995, I took Tom to the hospital to relieve his severe pain. He was diagnosed with lung cancer, which had metastasized to his bones, and he never came home.

Throughout his life, in his youth and as he aged, Tom was often ill. He was fragile. As a child, he spent considerable time lying on couches instead of sitting in school or playing outside with friends. This afforded him much time alone and the leisure to read, which he loved to do; and he praised his teachers of English and literature who had influenced his tastes. About ten years before he died, though, he gave up on fiction and read history unless I recommended some new writer or new work that I had especially enjoyed. He had an unusually sensitive nature, and beautiful poetry could move him to tears. He wrote his own good poetry,

 Human Incompetence

too. Each chapter of our book *Thinking Metric* starts with a fitting poem by Tom. He also had a great sense of humor, which the family and I so miss. As I said earlier, Tom was stamped 4F during World War II because of his recurring bone disease. And maybe that was fortunate, because he was so accident-prone and broke so many of his bones, even including one in his neck, that he probably would not have survived warfare.

*An evening with Friends in South Africa, 1983 (Marilyn and Tom under the zebra skin)*

Between us, Tom and I had eleven children. Tom had four children with Elizabeth (Betty) Battle:

- Kathleen Gilbert Hugh, born in November 1953

- Micah Gilbert, born in June 1955

- Sarah Gilbert Fox, born May 1958

- Adam Gilbert, born August 1960

He had one child with S.K. Dunn:

- Jessie Dunn-Gilbert Dikel, born August 1961

He had two children with Marilyn Bender:

- Roby Gilbert, born February 1966

- Eve Gilbert, born October 1967

He also had four step-children, Marilyn's children with Charles Ferster:

- Bill Ferster, born February 1956

- Andrea Ferster, born February 1958

- Sam Ferster, born August 1959

- Warren Ferster, born May 1961

Besides the children, Tom had eight grandchildren and four step-grandchildren. Quite a gang! We weren't always up to the job, especially the stepparenting part. I remember when my son hooked a phone setup to Tom's daughter's room so they could communicate late at night—or when his son and my son dangled our son out the second-floor window of our New York apartment. There were also many good laughs, as when my son bought Abbie Hoffman's *Steal This Book* and put it on the coffee table—and Tom walked by and stole it. Now, the children get together with or without me. Last October, the "Gilbert Girls" met at my house on Bainbridge Island. We plan to make it an annual affair, and perhaps include some who didn't make the first meeting.

 Human Incompetence

Tom died on September 25, 1995, at the age of sixty-seven. He never heard the OJ verdict. He never knew about 9/11. He never saw the inauguration of the first black man to be president. And because he never experienced the reign of George W. Bush, he never learned that it did matter who was president.

*Last picture of Tom Gilbert. Taken in June 1995, at his son Roby's wedding*

The Autobiography

For Marilyn Gilbert

*Who has been everything to me*

# CHAPTER ONE
## THE DEFEAT OF THE SIGMOID COLON

I'll begin by telling you where this chapter's title came from—it's a perfectly sensible one as you will soon agree. You see, when I entered graduate school in psychology at the University of Tennessee in Knoxville back in 1949, I could afford it only because there was a special job open to students who would pledge to enter clinical psychology. For thirty-nine hours a week I could work with World War II veterans, counseling them on their adjustment problems, and take all the courses I wanted. Since I was a newlywed, I needed the money, and scholarships were not available to me, what with the poor grades in my background. And this paid much better than scholarships. I wasn't particularly

 Human Incompetence

interested in clinical psychology, but so what. By now I had become an excellent student and was capable of pursuing anything in the field.

It may help you to visualize Knoxville back in those days. It was the gateway to the beautiful Smoky Mountains, but it wasn't so beautiful itself. Some well-known writer of the time called it the ugliest and most undesirable town in America, and he had a few things to support him. In those days, the air was so filled with soot that you had to change your linens daily; we smokers boasted that cigarettes helped clear our lungs. And the town had no special historic pride. Though occupied by Southerners, the main statue downtown is dedicated to the Union Army. Why? The hillbillies had never believed in secession; not because they held to noble principles, but because they owned no slaves. The nearest descendant of slaves may have been hundreds of miles away.

The university sat on one of the many hills surrounding the gullies of the town. On one side of "The Hill" sat the clinic of the Psychology Department, where those who presumed to hold the secrets of our inner workings held classes to teach the younger of us what those secrets were. At first I was curious about these strange ideas, Freud and all that. But pretty soon, as we branched beyond the great master, it all began to smell pretty dumb to me. I've long forgotten many of the names and particulars of most of the personality theorists we read, but I do remember writing on tests such nonsense as, ". . . and finally, the creative man returns triumphantly to the womb." And believe me, I always received A+ on my tests, and I was assured that the nonsense we were studying was standard for colleges across the nation.

But the work wasn't very demanding. A veteran/patient would complain that he was feeling awful, which made no demands upon the therapist who was only supposed to make a show of listening sympathetically: no problem unless you were suffering from a hangover. And, of course, you soon caught on; these VA patients could receive full benefits only if they made a show of their mental suffering.

So a couple of years passed with some time off working in a VA mental hospital in Memphis. Then came the final exam for the most advanced course in personality theory. I was prepared for anything the prof might ask but this:

"Your final exam will be to write your own personality theory!"

I was stumped only for a moment. By this time I so hated clinical psychology that I was prepared to flunk it; I was also qualifying in psychometrics—you know, testing and all that. So I lit into the exercise with joy.

First I stated with great certainty that all human motives stem from the experience of the fetus in its second trimester. Just as it becomes used to the great comfort of the gravid womb, it begins to sense a problem. Something is pressing upon it and then going away. Comfort is being totally compromised. Naturally, the tiny fetus does not realize that the source of this threat is the filling and emptying of the sigmoid colon—you know, the bottom part of the large intestine which is shaped sort of like an "S."

From that point on, the fetus tries to come to grips with

Human Incompetence

this disturbance, but just as it senses its power growing, it is ejected from the warmth of the womb out into the horror of the real world, no longer able to practice its first skills in facing conflict. From that point on, all of us are unconsciously driven by the urge to return to the womb to defeat the sigmoid colon.

*Tom in college*

Of course, this explains sex, both normal and deviant. And I went on to derive explanations for all manner of other human characteristics. I have forgotten most of my ingenious conclusions, but I do remember that tonic water was all the rage in those days. Its name? You guessed it. Hadacol. "Had a colon."

As I handed in the paper, I reasoned that if the prof wanted to flunk me, to hell with it. But I was not prepared for what happened next. Flunk me? Indeed not. He marked my paper with A+++ and called all the other students together—some ten of them—and required them all to read it. He also made copies and distributed them among the local psychiatric community. The psychiatrists had to meet me; they wanted me to talk to their local group, and everyone urged me to publish my brilliant insights. Wow! The air in Knoxville had never been so polluted.

I was soon awarded my Ph.D. and never again gave a thought to personality theories or clinical psychology.

One fellow student who also came to detest the nonsense did not fare as well as I. When asked on his oral exams what he would do with a patient of such and such symptoms, he said, "I would shave my head and reflect on my feelings." He went on to become a successful surgeon in New York City.

As a postscript, a gastroenterologist recently examined the inside of my sigmoid colon. I asked him if the colon did press upon the gravid womb. "I think so, but I'm not certain."

## Untitled Poem

*True learning's a moss that grows in darkened places*
*In gardens of wart and lichen, a growth antique*
*And constant. It is no flower of happy faces,*
*No profligate that spends its lust to seek*
*The sun and husbands too late to last through dread*
*Hibernal nights; But nourished by our tears,*
*It forms a pallet beneath an anguished head—*
*An herb enriched with potions for the years.*

*So when, my child, you seek this book to measure,*
*You ought not gauge its wit nor mental power,*
*Nor plumb its depth, nor draft its dizzy pleasure;*
*But rather calibrate its lingering hour,*
*And by this timeless scale you'll be assured*
*That through your wearying winter it has endured.*

*Tom Gilbert*

Human Incompetence

## The Youngest Member of the Rorschach Research Exchange

Not all of my Ph.D. studies left such negative impressions. One of my profs, the late E.E. (Ted) Cureton, was considered by many in those days as the psychometrician's psychometrician. I became an expert on every kind of test you can imagine, including the Rorschach Ink Blots (which I secretly still adore—shhh. I once became the youngest member of the Rorschach Research Exchange). And I learned all about the IQ tests, achievement measures, personality tests, and the like.

At some point I received Cureton's support for studying the correlation between IQ and work performance. Knoxville was surrounded by industry in the new atomic age. Oak Ridge was only a few miles west. Ted set me up and I made my first entry into the workplace. The manager assigned to help me was as excited as I was. And then the first disappointment hit me. What do you think I found? Go into the workplace today if you want to observe the same phenomenon. There were no measurements of job performance. To obtain correlations, we had to set up our own measures, which excited the manager even more.

## Job Description

*If all were artists and artisans none,*
*There'd be poems without paper for printing them on;*
*A footless Phidias, the fundament gone;*
*Songs and no cymbals to sing them with;*

Confessions of a Psychologist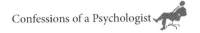

*Plays seeking podia to plot them upon;*
*And canvas without cradles for curbing the myth.*

*Then, frameless painted works of heart*
*Would mix and meld and overheat*
*And, streaming through the somber street,*
*Consume the world in angry art.*

*Tom Gilbert*

Then what do you think I found? You guessed it. No correlation. So I moved to other jobs. Again we set up measures because none existed. And what do you think we found? Yep. No correlations. But this was back in 1950. Today, all that has changed. These days hundreds of studies report that IQ scores account for as much as 50 percent of the variance in job performance. What has changed?

The job measurement situation certainly hasn't changed—they still barely exist, and when they do, they rarely are the correct measures. How about the IQ tests? Nope, they've hardly changed—I used the Wechsler-Bellevue, still considered a standard. The jobs certainly haven't changed. So, what has?

Ah! In my day, I may have been the only person studying such relationships. And I had nothing much riding on my work. Jobs were opening up for Ph.D.'s. I had a lot of stuff for my Ph.D. thesis that had nothing to do with these

 Human Incompetence

curiosity-driven studies. What has changed?

Well, these days, hundreds of folks are struggling to get their Ph.D.s to compete for the few jobs that go around. Negative results do not fare very well in the research community, so positive results are needed to build up a publication record. The number of positive results has been steadily growing over the years since I had my failures.

> *Meter what your Fancy will;*
> *She'll doubtless conjour Joyous Hour.*
> *My love—she's more audacious still—*
> *Challenged me to fathom Flower.*
>
> *But I have measured everywhere,*
> *By magnum at night, by minin at morn,*
> *And neither Rose nor Vine compare*
> *As cadent, sharp, and true as Thorn.*
>
> **Tom Gilbert**

Indeed, some of the positive results are very strange. One study reports a correlation between downloading a conveyer belt and IQ. Could that be? I know a lot about downloading conveyer belts, and this has struck me. There is virtually no variation in the performance of downloaders. Indeed, many conveyer belts are driven at a fixed speed. So how can one find a correlation? If you are not sure of what I mean, think of going into the woods to study the correlation between the height and breadth of trees. But if all the trees are of the same height, you'll never find a correlation.

Another study reports a high correlation between bank tellers' performance and IQ. But, again, where is the variation? The top tellers close out their cages at the end of the day with close to 99 percent accuracy. This is a bit better than the average, which is close to 98 percent. To find a correlation, you would need a detector of microscopic precision. So where do these correlations come from?

I think I have discovered where. Far be it from me to suggest that psychologists might fudge their findings as medical and pharmaceutical scientists are suspected of doing. Pharmaceutical research standards are very tough. Here are just a couple of them: The studies must be "doubleblind," meaning that neither the subjects nor the experimenters know who is receiving what drug or placebo or who scores low or high. And strict standards are held for reproducibility. No matter what exciting thing you find, a disinterested party must be able to get the same findings.

Even with such strict standards in place, the medical research community has to be carefully scrutinized for honesty—so much rides on their findings, including jobs. Frequently we are reading about a major scandal in this research community; but few are aware of how widespread the problem really is. Over the years I have counted several pharmaceutical companies as my clients, and the problem is much worse than it appears to the public.

 Human Incompetence

Of course, one psychologist angrily retorted that psychologists don't cheat. Well, perhaps not. But I am not aware that double-blind and other standards have ever been adhered to.

So where do these performance measures come from that did not exist back in my day? Aha! We now have something called (laughingly, I would hope) "performance appraisal" measures. Annually, supervisors are asked to rate their employees on a whole list of behavior traits, such as "creativity," "initiative," "punctuality," and the like. Typically, a five-point rating scale is used. Rarely are people rated on what they actually "created" or "initiated."

I recall once receiving a very low rating on "initiative." Let me tell you about it. A client wanted me to observe some of the management training it had in place, but I had to sneak in and pretend to be just another manager. In the middle of some exercise, one of the instructors interrupted us and said we must get up to the fourth floor. We all gathered at the elevator on the first floor and waited. But the elevator was slow in coming. Finally, one of the trainees ran up the steps, shortly followed by others. I noticed the instructor was marking something on her chart. The elevator never came, and I was the last to walk up the stairs. Later I learned what the exercise was all about. The first man to run up the steps was rated 5 on the initiative scale, and I received the lowest rating of 1—no initiative. Alas, the scale did not reveal that I had recently undergone spinal surgery and found that climbing steps was an initiative hardly worth the price.

Confessions of a Psychologist

## "You can't write seven pages without making an error"

Well back to the sooty hills of Knoxville to finish off my graduate education. When I moaned to Ted Cureton that our tests didn't seem to correlate with anything very well, he nodded sadly and said, "Well, some day, perhaps."

Ted also gave us a first-day final exam in our most advanced course in psychometrics. There is a psychometric measure called the *reliability coefficient*, which is very easy to understand. "How well does a test correlate with itself when you administer it again?" is one such index. If it is perfectly reliable (physical scientists would call this "precise"), you will get a correlation of +1.00—meaning a 100-percent repeat of the first scores.

The man who invented the reliability coefficient was known by many as the "father of psychometrics." On the first day of our final course, Ted told the four of us that the most famous paper written by Truman Kelly was called the "Reliability Coefficient" (*Psychological Bulletin* 1941). He had reserved four copies in the library for us. Our problem was to find the error Kelly had made. Wow, what a simple assignment!

Until we got to the library and found that we couldn't even understand the paper let alone find an error in it.

Late in the semester, having studied about everything in psychometrics, we began to understand Kelly's seven-page paper. But we couldn't yet find an error. Finally, at semester's end we reported our errors. There was some overlap,

Human Incompetence

but the errors varied from maybe five to nine. Of course, we competitive lads had never discussed them with one another.

Ted gave us all an A except me. He gave me an A+. At dinner later, Ted explained my A+ to the others. "I found Tom's error the most interesting," he said. No one argued.

"Ted," which error did you have in mind," we all wanted to know. Oh, I never saw an error. I just know you can't write seven pages without making an error." What an education!

Many years later I took Ted and his wife to dinner. He had retired but he recalled our test. "Ted, I've forgotten. Why did you put the plus on my A?" "Oh, that was a mistake. Yours wasn't an error." And he explained, "Will that change my grade, Ted?" I asked teasingly. "Oh no; it's a simple matter of reliability."

## Cureton's Law

I think the most memorable lesson we received from Cureton had to do with the meaning of differences between groups of people on test measurements. In our day there was a burgeoning of studies designed to prove that whites were a little brighter than blacks, for example. Of course, none of these studies could pass the pharmaceutical research standards, but they kept coming forth at such a rate that Cureton finally commented on them. He drew two large normal distributions on the blackboard, slightly overlapping.

"Notice what would be true even if the tiny difference between these groups were for real. Nevertheless, the variance within groups is still much larger than the variance between groups."

Later, when I described this phenomenon to a class, one of my students agonized over the meaning.

"You mean to say that even if whites were brighter than blacks on the whole, 48 percent of the blacks would still be brighter than the whites?"

"Forty-nine percent," I corrected him.

I think this should be known as Cureton's Law: "The variance *within* groups is greater than the variance *between* groups."

## Mendel Flunked Genetics, Didn't He?

When I entered the University of South Carolina in 1944, I was the first person in my family ever to attend a college. It was my mother's wish that I go to medical school, or into law or business as alternatives. But no three subjects could have interested me less. However, penny-pinching Mom was paying, so I dutifully studied the sciences. I was completely unable to comprehend physics, but a little cheating got me through with a C.

But I sailed through courses in philosophy and psychology—except for one, a course called *Applied Psychology*. I felt that I had probably squeaked by with a C and could graduate; but the professor gave me an F. I naturally went

 Human Incompetence

to him pleading for help. He made a deal with me.

"Tommy, you are considered a very good psychology student, so here is what I am going to do. Promise me that you will never, never pay further attention to applied psychology and I will change your grade to a C. It just isn't your thing."

On my knees I swore off forever. Today, I don't feel very guilty about my oath: at the time I gave it in good faith.

Although my average grades as I graduated were a flat C, I made very few actual Cs. Mostly, I earned As or Ds, depending upon my interest. I earned a D in art appreciation simply because I wondered how they could teach such stuff in a college. Biology was simple and fascinating, and the professor begged me to make a career of it. But there was nothing like philosophy, and I would have pursued that to the end had I not encountered one of the great influences of my life, the late James R. Simmons. I'll have to tell you a little about him.

## "...A he-man wants his atheism!"

When I first met Jim he was just entering the University of South Carolina and we were taking a course in English drama, I as a senior and he as a freshman, though he was about twelve years older than I. Jim worked as a printer and had difficulties communicating with the customers in his sister's shop. His sister recommended that he take a course in public speaking on his GI bill, one taught by Chris Christopherson, the university's drama professor. Jim seemed terribly shy and kept asking for advice about this and that—he sat next to me in class. By my senior year I

## Confessions of a Psychologist

was able to give confident advice to anyone. Christopherson had talked Jim into enrolling as a fulltime student—"What should I major in?" That was easy—"Philosophy, of course, and here are all the superb reasons."

Next, I got Jim a job as a fellow clerk in the University Personnel (Testing) Bureau. Professor McCall would employ anyone who scored very high on the achievement tests of the day. Simmons had been a guinea pig for a freshman take of the Graduate Record Examination and had the highest scores McCall had ever seen. He also scored at the very top of the college entrance exams.

One of the great pleasures of working for "Prof," as we called McCall, is that he encouraged his people to argue and discuss things intellectual—as long as their hands kept moving as they collated test papers or whatever. So at first I was pontificating to the naive Mr. Simmons about the fundamental truths of philosophy. If he queried me a bit, I quickly set him straight and went on to the next truth—you see, I was preparing for a career as a college professor. This went on for two or three months, and now I will jump ahead to tell you what happened.

In two years, Jim Simmons had left his stammering ways and won the international debate championship—I think Oxford was his last victim. And then he won first place in the international extemporaneous speech contest. He was a tall, skinny guy with a thin cynical face and the most superb sneer I have ever seen. Pretty soon he debated every smug argument I put to him, usually leaving me crushed. "Ah ha! Jerked your God out from under you, did I?" Gradually, my

 Human Incompetence

certainties turned to humble questions: "Jim, what do you think of this?"

He was no rigid intellectual, though. I remember a group of us discussing religious beliefs. The issue came up about the comparative advantages of agnosticism and atheism. One of the young freshmen asked me what I thought. By this time, Jim was teaching the basic course in logic, and I was eager to impress him.

"Strictly speaking," I pontificated, "atheism is an illogical stance. It violates a basic rule: the 'fallacy of the affirmation of the negative.' You can not prove a negative, though you can fail to prove a positive statement. Therefore, agnosticism is the preferred stance logically. Isn't that right, Jim?" I was eagerly seeking his approval.

"Bullshit! Agnosticism is for sissies; a he-man wants his atheism!"

The issue of agnosticism and atheism came up once again many years later in my life in Mendham, N.J. Our youngest son, Rob, 11 at the time, had just starred in an adult version of the musical "Oliver!" He was Oliver! Of course, we attended every presentation and the final cast party on the grounds of the Episcopal minister's house. At about two in the morning, I went to collect Rob — who was listening carefully to the preacher, who was explaining the difference between agnosticism and atheism to a young couple.

This was not Rob's first encounter with atheism. At age six he was in a private school in New York. As the Jewish holidays approached, the teacher wanted to know which students would be absent. Rob's mother is Jewish, and when the teacher asked the Jewish kids to raise their hands, Rob only nudged his up.

"Rob, are you Jewish or not?"

"I'm half Jewish and half atheist."

As we got into our car for the seven-mile drive home, Rob sat up close behind me and wanted me to continue the religious discussion. To the best of my knowledge, little Eve, age nine, was asleep in the back of the station wagon.

My pompous lecturing had satisfied Rob as we arrived home. But as I entered our drive, a little voice came from the back.

"Daddy, what do you call people who just don't care?"

The champagne from the reception heightened the challenge. By breakfast time I had found the answer in my library. There was a group who wandered the Near East about Christ's time who called themselves "securitans—secure without care." It struck me that Eve may have uncovered a great hidden movement. Nearly everyone I tell this story to allows as how they are securitans at heart.

Her mother: "Eve, now that you are the popess of the securitans, what do you plan to do?"

"Make Daddy stop smoking."

"But I thought you didn't care!"

"I don't care about God, but I do care about Daddy."

 Human Incompetence

## "Psychology, camp follower of science"

But back to Jim Simmons. Jim went on to Columbia University and earned a Ph.D. in philosophy, taught a while in Utah, then returned to South Carolina to teach philosophy. "I owe it to my own people. I have never sold out to them," he explained. But Jim was responsible for my giving up a career in philosophy. "True philosophers never had Ph.D.s in the subject," he argued. "Descartes died shortly after he began teaching." "But, how about John Dewey?" "He was a teacher of philosophy. His sidekick, William Bentley, who really did it all, didn't have a Ph.D. Nor did Isaac Newton."

"Newton? He was a physicist!"

"To hell he was. He was a philosopher who invented physics. That is what real philosophers do; they invent things."

"But, Jim. What shall I do? I have a Ph.D. in psychology, and your thesis at Columbia was entitled, 'Can There Ever Be A Psychology?' And your answer was pretty much, 'No.'"

"Well, this is your chance to make a science of it! Perhaps I am wrong. But first you must solve the mind-body problem!" I'll come back to that later.

Jim recounted his Columbia professor's reading his thesis and looking up to say, "Jim? What about after images?" And I recall Jim remarking, when he learned that the Psychology Department was in the old physics building, "Psychology! Camp follower of science!"

## Confessions of a Psychologist

The mind-body problem! Not many teachers of philosophy take this too seriously anymore, but I did. I'll try to keep it simple for you. You've heard the question, "If a tree falls in the forest when no one is around, does it make a noise?" Well, that is sort of a "mind-body" question. The old philosophers used to wonder, "Is all the world simply a mental image, or is the mind merely a reflection of the physical world?" You may not be consumed by this question, but I certainly was, particularly after Jim said I had to solve the problem before there could ever be a psychology. You may feel more like the guy who once wrote disdainfully, "Never mind? No matter. No matter? Never mind."

But I struggled mightily with the issue—and came up with a solution, one that was, as Jim required, "parsimonious, elegant, and useful." (You'll hear more about this later.) Jim listened carefully and said, "Well, you've done it; but how are you going to sell it?" (If you want to read the solution, see Chapter 11 in my book, *Human Competence*, Salem, Mass, HRD Press, Second Edition, 1994.)—or should it be in this book?

Jim died of a massive heart attack in the 1950s, but he had a huge effect on me.

### Know your animal!

I entered teaching at the University of Georgia in 1952, a brash young assistant professor who knew everything. I was twenty-five and everyone else there were old fogies, though nice enough people. One of them was to have a great influence on me in many subtle ways. This was Bill James, an

 Human Incompetence

old curmudgeon, maybe twenty years my senior. Bill ran the animal laboratory and was a behaviorist, largely of a Pavlovian bent. Despite his growly demeanor, he had the loveliest wife and two daughters I think I have ever known. I once asked him if he practiced behavioral principles at home. "I don't know enough about that stuff—and I'm not going to treat my family like dogs!"

A few years later I had gotten to know Charles Ferster quite well. Charley was my wife Marilyn's first husband. It was a disastrous marriage, though they had four fine children. Ferster had been B.F. Skinner's right-hand man and ran his pigeon laboratories. Together, they had published a landmark book, *Schedules of Reinforcement,* Ferster and Skinner, 1957.

Their marriage was disastrous for a number of reasons, but not the least because of Charley's agenda for raising the children by behavioral principles; or as Bill would say, "to treat them like dogs." At one point I was attending a psychological convention in Atlanta, and Charley grabbed me and squirreled us away in the privacy of the rear of a bar. I had already had my first kid, and the second was close to being born. Charley wanted to learn from me how I was applying Skinner's principles in raising them. Of course, I was following Bill's advice, but ever one to be impressive, I answered all of Charley's questions with great confidence as he avidly took notes.

But then we were interrupted. Bill James walked into the bar and joined our table uninvited and listened to me pontificate without comment.

Charley was visibly irritated by the interruption, but

proceeded to ignore Bill. Finally, he came to the first question that ever stumped me.

"Tom, what do you do if the child has a temper tantrum?"

At a complete loss for words for once in my life, I turned to Bill for help. "You ever have this problem?" remembering Bill's lovely daughters. "Hell yes. Once."

"Well, Bill, what did you do about it?" His answer couldn't have annoyed Charley more.

"I put a foot stool over them. That'll do it every time."

Recently, Marilyn and I have been raising our granddaughter, Georgia, who has had a tantrum or two of late. Annoyed, Marilyn has turned to me. "You're the psychologist; what should we do?" As I write this, I don't recall the solution. But our hassock is much too large, so we'll have to get a smaller one.

But I believe that Bill's biggest influence came one day when I was training a cat in a large Skinner box. In Skinner boxes you can train animals to press bars or peck windows to get food.

"Bet you can't train a possum in there," Bill challenged. "Bet you ten bucks."

My cat-sized Skinner box was made of tin and had a food pan at one end with a bar mounted well above it. I took Bill's challenge immediately. I would teach him how universal Skinner's principles were! I had better—ten bucks was a lot of dough back in those days.

 Human Incompetence

But somehow I wasn't succeeding. I couldn't get the possum to go up that tin wall to the bar, even by the most careful efforts to shape him. And I couldn't figure out why not.

Finally I paid up. "What did I do wrong, Bill?"

"Possums don't have a neural connection between their brains and their hind legs. This should be a lesson to you. Know your animal!"

## "You're not extinguishing him; you are reinforcing him!"

Poor Charley Ferster. All he wanted to do was establish the universality of Skinner's science. But he could have used Bill's lesson, slightly reworded: "know your child." This memory should be useful to any of you who are anxious to rush home and apply behavior principles to your family.

Ferster was without argument the world's foremost expert on the reinforcement and extinction of animals. Read the book, *Schedules of Reinforcement*, and see all the work he had done—he could tell you exactly how extinction (meaning the withdrawal of rewards) would affect the animals, depending upon their previous history of reinforcement. If they were used to getting a reward every time they pressed a bar or pecked a window, they would become quite erratic and upset when the reward was no longer forthcoming. But if they were used to being paid off randomly, just every once in a while, they would work like hell for you when no rewards were forthcoming. Other histories of reinforcement produced other patterns of behavior during extinction.

Confessions of a Psychologist

Ferster knew them all.

No one in the world could describe the effects of extinction like Ferster. No one had ever studied it in such detail.

So how did he apply this knowledge at home?

Once, when son Sammy was about two, he took to jumping off the couch, over and over again. This began to annoy Charley, so he decided to practice his skills as an extinguisher. Without comment, Charley grabbed Sammy each time he leaped from the couch. And he grabbed and he grabbed and he grabbed, showing great patience. But it wasn't working. Sam took to jumping as never before. Until Marilyn noticed.

"You are not extinguishing him! You are reinforcing him. He loves your attention."

## Knowledge Maps

*On charts of knowledge the roads run west*
*Through thicket, fen, and squalid town,*
*And weary travelers wanting rest*
*Will search for coves that can't be found;*
*For every quarry one would quest*
*Lies safely hidden underground.*
*But once my Love did dwell along*
*These paths, so I will delve for what she knows,*
*For she can tune the catbird's song*
*And ameliorate the rose.*

*Tom Gilbert*

23

 Human Incompetence

## Someone already has begun reconstructing psychology

Another of my influences was an old friend and fellow student by the name of Joe Hammock. Joe influenced me in many ways, but one way in particular. When I was assuming that all human deficiencies in the place of work were indications that training was necessary, Joe pointed out that there were plenty of problems that were caused by other things—a simple idea, but not one that readily occurs to those attached to the importance of training. So I began to make a formal distinction between what I called deficiencies of knowledge and deficiencies of execution. I even gave the distinction a fancy abbreviation: De vs. Dk, I called them. It's not much on the surface, but it caught on big in the training community.

### The Laws Of Training

*Warns the Law of the malleable mind:*
*"Do not linger, don't digress."*
*With the ease it grasps good sense*
*The human head absorbs a mess.*

*Mental geography's relentless rule*
*Our toughest teachers were never taught:*
*The shortest route is the straightest point*
*Between the distant lines of thought.*
*The keystone to the human head,*
*From Kew to Constantinople:*
*To optimize the worth of training,*
*Render it as rare as the golden opal.*

*Tom Gilbert*

But Joe's biggest influence on me came in another way. I remember telling him about how I had studied all there was to be known about human learning, and that the field was a complete eclectic mess. I felt that I was going have to do with it what Descartes had done with philosophy, start all over from scratch. But Joe said that someone had already done that, an interesting fellow by the name of B.F. Skinner.

Oh, I had heard of Skinner, and he was about the only psychologist I had never studied. Someone had told me that he was stuck on pigeons pecking in a box, and you certainly can't learn much about learning that way. They also called him a "logical positivist," an insulting philosophical term for many of us and the correct answer on the tests I used to take. But just before going my own way to reconstruct all of psychology, I followed Joe's advice and sat down with Skinner's first book, *The Behavior of Organisms*. (New York: Appleton-Century-Crofts, 1938). This was about 1955, and I was a young assistant professor at the University of Georgia. I've never been the same since.

# CHAPTER TWO
## INSURRECTION: 1950 STYLE
### A PERSONAL HISTORY
## DISCOVERY OF ENGINEEERING

Even at the age of ten, I had an enthusiasm for teaching. My six-year-old brother did not take too well to my urgent attempts to instruct him in addition, yet the desire to teach has never been a disease curable by unresponsive students; I am certain there are plenty of professors who would lecture away for many minutes even after their students had stealthfully sneaked away from their classrooms. Fifteen years later, I was still at it, not only professing myself, but playing the scientist, studying learning in a laboratory at the University of Georgia, in musty old Meig's Hall.

But the desire to comprehend teaching and learning was not my only disease; I also had a severe case of obsessive-compulsive neurosis: lawfulitis. I had abandoned clinical psychology when I finally realized I could never write a Freudian equation. So severe was my compulsion to find lawfulness in matters of the mind that even the "laws" of learning left me unsatisfied. There weren't many of them that seemed very reliable. That is why I settled on the Kjerstadt-Robinson "law" as the basis for my research. Both Kjerstadt and Robinson, in independent studies, had proclaimed that "the rate of learning is independent of the amount of the material to be learned." That is, if you memorize a short list of nonsense syllables in a 20-percent segment of the total learning time, you will do the same for a long list.

Now, admittedly, Kjerstadt's and Robinson's law is somewhat less exciting than cosmology or $E=mc^2$. But, nonetheless, it did promise order in a field all too disorderly. Testing the law had also helped me get a rather painless Ph.D. and provided a third confirmation. Then, I decided to delve more deeply into it, although I would have to reproduce the original findings of Kjerstadt and Robinson. I did this by subjecting student volunteers to the memorization of long lists of nonsense syllables (e.g., PUV, XIK, YAL, etc.) and paired associates (e.g., Mate – carp, sand – live, etc.).

In conducting such experiments, the unfamiliar reader should know, one proceeds by carefully standardized methods toward a precisely calculated statistical result. Some students will learn faster than others, of course, but nearly two centuries ago Carl Friedrich Gauss gave us the "bell-shaped curve" techniques to deal with this. Statistics are based on averages and deviations, and the deviations from the average normally follow a predictable pattern. If the average student

 Human Incompetence

takes two hours to learn my list of paired associates, I can reasonably predict the number who will master the list in as little time as thirty minutes (not many, of course). Occasionally, however, in such experiments, a "subject" (as the students are called, presumably to keep them from sounding too subjective) will deviate so far from the average that you judge him "unrepresentative." An example would be a student who simply was unable to learn the list at all, after many hours. But now I'm going to tell you the story of a "subject" who deviated in the opposite direction and set me on the course of becoming an Educational Revolutionist.

Meig's Hall was built in the early nineteenth century, and its basement smelled like it had last been dusted right after its construction. Robby Bender, our subject, enters the Learning Laboratory with a reluctant shuffle—a student volunteer under the incentive of extra credit. The atmosphere of the room is thick with the quiet throb of earnest science, shades of Wilhelm Wundt and E.B. Tichener. Robby is seated before a strange device called a memory drum, a black, round instrument with a small aperture on its cylindrical surface. The aperture is designed to display a word or syllable, one at a time, carefully timed in sequence, so learning can take its lawful form.

Robby is given the standardized instruction. He is unlucky; he has drawn one of the longer lists—one hundred paired associates. His task is to learn to say "carp" whenever the word "mate" shows in the window—and ninety-nine more of these carefully chosen pairs. The first time through, of course, he is only expected to pay close attention and learn what he can. Then, on the second round, he is required to look at a stimulus word (e.g., "mate") and try to recall its paired associate, making the response "carp," if he can remember it. This is dull work, but Robby doesn't realize how much more boring it is to me than to him. Only the highest dedication to scientific truth (and the advancement to an associate professorship) can sustain the endless repetitions of the memory drum I anticipated would be necessary before Robby masters this task, expected to be a matter of hours, not minutes.

On the first time through the list, Robby pays strict attention. Then he is instructed to pronounce out loud the anticipated word when he sees the first of the pair. Pencil ready, I begin to record his right and wrong answers—except, there are no wrong ones. Robby masters the list in one trial. Of course I'm impressed, but question Robby Bender very carefully. Could he have violated my system of security and learned the list earlier? He's only a B student, but I can find no other basis for suspecting him.

Now, Robby is scheduled to return tomorrow to learn another list, but I can already see that he threatens to seriously muddle the statistics I will be using. He clearly is "unrepresentative," and so I inform him that he is no longer needed. He looks gravely disappointed, but leaves without protest. I'll not have the Kjerstad and Robinson law violated by an oddity.

"He's probably using some sort of cheap memory trick," I think to myself. The next day after class, Robby lingers until everyone else had left. "What can I do for you, Kiddo?" I ask lightheartedly, sensing his disappointment.

 Human Incompetence

"Professor, why did you throw me out of the experiment?" he nearly whines.

His response to my long, scholarly explanation—bell-shaped curves, statistical representation, and so on—makes it clear he is unconvinced. "But I thought you wanted to learn how people learn—and since I learned that stuff better than anyone else, why don't you throw the others out and study me instead?"

The question hits me like a sledgehammer.

Meig's Hall, I understand, still stands and in its ancient room, the fiftieth generation of students will soon be learning its catechisms in much the same manner as all those who passed through those portals before. I wonder if the old memory drum is still clicking away at this very moment, twenty-five years later.

Robby Bender, as I have called him, having forgotten his name, was given a flustered and defensive answer to his parting question. I wasn't about to let a student and an oddity tell me, a scientist and professor, how to study learning.

But that night, Robby's question wouldn't go away. How in the hell did he learn that list in one trial? I hadn't the faintest idea. Probably a memory trick, but what kind? It painfully dawned on me. Here I was, the state of Georgia's expert on the science of learning, and I couldn't learn a list of words nearly as well as a B student. Worse than that, I began to admit to myself, I didn't even know how to teach a course on learning very well. Why, I couldn't even train a dog. I abandoned those experiments, because my heart wasn't in them. I was a fraud, and I knew it. I wish I knew what happened to Robby Bender; I haven't been the same since his question. Unfortunately, I was still not ready to grasp the full significance of Robby's question for me. Do you want to study behavior as a scientist, or do you want to engineer it? They are quite different things, you know (see Gilbert, *Human Competence*, 1978, pp. 2 – 5). First, I had to get science out of my system.

Shortly after Robby's devastating question, I set out to reconstruct the science of learning from the ground up, Cartesian style. I would admit I knew nothing and start from there. As I explained this plan to Joe Hammock, a friend who had been a fellow graduate student, he suggested I read B.F. Skinner's book *The Behavior of Organisms* (1933). Skinner, he explained, had essentially done the same thing and had gone some distance with it.

The name Skinner, of course, was familiar. Hadn't I been an excellent graduate student and supplied the requisite paired associates to get my degree? Sure—Skinner: operant conditioning, positivism, and reflexes. The guy who didn't believe in learning theories, which is all I wanted to believe in. But Joe had often put me on to something interesting.

I completed the first three chapters of *The Behavior of Organisms* a religious convert. Not since *Fanny Hill* had I been so aroused. Here was lawfulness and order. Here was methodology. And here was a guy who could actually train an animal, not just watch it run down mazes. Indeed, he could train animals better than expert animal trainers. At the very least, now my mother couldn't say, "You teach

learning, and can't even train a dog?"

Before long, the basement of Meig's Hall was filled with Skinner boxes and cages of animals of all sorts. Soon, I had found my research niche, very basic stuff: experiments on the ways of measuring operant behavior. Somewhere from the dimensions of the operant there might emerge the solution to the ancient mind-body problem. "If a tree falls in a forest when no one is around, does it make a noise?" "What is the fundamental difference between mind and matter?" (Skinner's pooh-poohing these questions didn't put me off them.) And sometime there might emerge from all these orderly graphs of behavior the *Great Educational Revolution*.

But I wasn't wholly satisfied. Still the sovereign state of Georgia's learning expert, and while I could now train animals, I couldn't train those human students of mine very well. So it wasn't long before I had a human Skinner box, complete with a bar to press (actually, a lever to pull), and all sorts of ways to present stimuli. My cumulative recorders were churning out data as I tried to think of ways to teach my subjects (they were still "subjects") something complicated, like algebra or operant conditioning. And I was about to be enlisted into the Great Educational Revolution.

The call came in the form of a paper, Skinner's "The Science of Learning and the Art of Teaching" (1954): The teaching machine had arrived. The Revolution was not only here, it was nearly over! Great joy will soon be heard throughout all the land.

Of course, there was still a small amount of work to be

done—only details, naturally, but details are important. The word had to be spread, teaching machines out into all the classrooms, studies to be reported, and the absolutely predictable success to be measured and trumpeted. I wanted to be in on that final push, even as a little foot soldier in Skinner's great army.

So I built my first teaching machine. A huge, ugly monster, full of relays, mechanical gadgetry, and buttons to push to allow for "branching"—individualized tracks through a learning program (from what I could tell, Dr. Skinner's teaching machine seems a mite too simple to satisfy my total enthusiasm).

But two of the small details I was faced with kept causing me trouble. First, the damned machine kept breaking down; and second, the only thing by way of introduction I had to put into the machine were test questions. They didn't seem to teach very well.

I solved the first problem by building several varieties of simpler machines—the simplest worked (my sharp, scientific eye soon noticed) as well as the most complex.

Now, the idea of the teaching machine, for those who haven't heard, is simple. You show some information in a window, ask for a response from a student (usually written onto paper through an aperture—shades of the memory drum?), and, most important, the response gets reinforced by letting the student see the right answer, just as the pigeon in the Skinner box gets his food pellet immediately upon pecking a key. Soon, to my surprise, I was arranging for all these things to happen even without a machine. I used a

 Human Incompetence

long manila folder with an aperture in it. The student could fold his teaching program into it, read the information, write in his answer, and slide the paper forward to see the correct answer. It was a little disappointing to see the teaching machine, this greatest of all historical concepts, reduced to a mere manila folder.

And then I noticed one of my students working with the materials without the manila folder. He simply put his hand over the answer, read the information, made his response, and slid the paper forward to get his reinforcement. And that worked just as well as all the other devices—in truth, it worked better.

My God! The Great Educational Revolution had vanished with no saving device and a list of test questions. That's exactly where I was before the Revolution, only worse: I was doing nothing more than teaching students to the test.

You might think that my despair would be suicidal in its dimensions. But, happily, teaching machines and revolutions were only a sideline—I was principally engaged in trying to prove that I was a scientist. Skinner read my work on the dimensions of the operant and invited me to Valhalla, then located in the basement of Memorial Hall at Harvard. After this invitation, we corresponded a bit and he learned of my interest in teaching machines. That winter, I faced a long, impatient summer before leaving the woods and entering the temples at Cambridge. What should I do to prepare myself for glory? Teaching machines were little on my mind—I was headed for the heights of real science.

But the Great Educational Revolution was on and there was

no stopping it. Publishers had learned about it, as had Wall Street—and Bell Telephone Laboratories.

Herb Jenkins, operant conditioner and pigeon trainer, labored away in the bowels of the great building in Murray Hill, New Jersey. Naturally, Herb kept his colleagues at Bell Labs cognizant of Skinner's work, and naturally, Bell Labs wanted to be right in there at the beginning of any new scientific revolutions. Why not start with the revolutionary general—ask Skinner himself to start a "Teaching Machine" Department at the Labs. Skinner remembered I had the summer off, and soon I was dispatched as his missionary, to bring the Revolution to The Bell System.

Just in case, I took along one of my machines, a ratchet affair with a lever that cost me $50 to build in Georgia. Surely, since Skinner still believes in teaching machines, there might be something in it. But by now, I was more interested in figuring out how to write the materials that went in the machine. At Georgia I had, with a friend, developed a "teaching machine" course in remedial algebra, and another to teach basic psychology (Skinner's style). These courses conformed to Skinner's pattern of developing a formation—instructional sentences with blanks to them—*frames*, they were called.

My mission at Bell Labs was simple. In three months, as a resident consulting "scientist," I was to write a set of principles of teaching machine instructions and help them start a department.

Thus emerged a document I called "A First Approximation to the Principles of Programmed Instruction." It was not

 Human Incompetence

published but was printed and spread around. There were no real principles in this paper, of course. As I recall, the offerings consisted of such blinding insights as "be certain the items of instruction are listed in the order in which you want to teach them," "break the items of instruction down into their smallest informational units," and "identify the stimuli and the responses you want students to make to them." And, in the style of biology, I carefully classified some thirteen different kinds of instructional frames—"augmenting frames" introduced new information; "dovetailing frames" related two pieces of information to each other; and so on.

By the end of the summer of 1952, I had two courses of programmed instruction being developed in the Bell System, and a manual of procedure for writing these materials. Imagine my dismay when, a year later, I found dozens of people following my manual as gospel. You see, when I left Bell Labs in September, I no longer believed a word of it.

Indeed, as I entered my new office at Harvard, I was nearly a complete cynic, believing the Revolution to be lost, and I quickly immersed myself into the "true science," the study of the dimensions of the operant. But the flames of insurrection were rapidly spreading. In Batchelder House, I shared an office with Sue Markle (then Meyer), Skinner's assistant, who daily plodded away at creating what seemed to be countless frames for children's arithmetic. Skinner and Jim Holland were beginning to plan a programmed textbook on operant conditioning. A publisher purchased my $50 teaching machine (and all rights to it) for $500, even over my protests that it was useless. Out in the boondocks, people like Jim Evans and Lloyd Homme would soon be putting together new programming principles ("ruleg-show a rule, then give an example; or "egrule," —show an example, and then give a rule). And Norman Crowder was advancing a "wholly new concept" in programmed instruction—the "branching" textbook, different from Skinner's "linear" programming. In this book, if you got one kind of wrong answer, you were directed back, say, thirteen pages. Another wrong answer and you were directed to the next page. A right answer would allow you to skip a few pages. Teaching machines proliferated. (At a convention, the teaching machines on display made the exhibit look like a museum of my failures at the University of Georgia.) I dug deeper into the "true science."

But the possibilities of the Revolution haunted me. The true source of its impending failure, it began to seem to me, was the lack of any fine, sophisticated order in the way we designed programs. "Break the information down into small steps, make sure the student makes a correct response (without overprompting), and then reinforce the response immediately." That was the whole theory—not very elegant. Great issues abound—who really first invented the manila folder "teaching machine"? who discovered "branching"? (Cain, I suspect, as he instructed his brother Abel.) Finally, "how do you really spell *programming*—one 'm' or two?"

Strangely, we operant conditioners, the greatest animal trainers, used far more sophisticated principles for teaching our animals than we did in programming materials. Why?

And then it occurred to me—we were going about it all

 Human Incompetence

wrong. We were approaching the subject of programming instruction as scientists might, not as engineers. People were already embarked on experiments to compare the effectiveness of one approach to another. And what do you learn if you discover that one lousy method of instruction is statistically superior to another, when Robby Bender (non-scientist) knew how to teach a list of the paired associates in one trial.

Indeed, as scientists, we knew a great deal more about learning than we had ever applied to teaching people. Most of the learning principles we were reasonably confident about never showed their way into the instructional programs.

And scientists, as I've said, approach nature to study it, not reshape it. They develop a method and follow it wherever it might lead them. But there is another view—a view that says there is much more known than we've ever applied. Education is not a natural science. The proper view is to approach it as something to shape, not to study, using whatever relevant knowledge we now have and adopting or designing whatever methods might be useful to reaching the ends we desire. The question is not "how well do children learn to read by this or that method?" but "how best do we teach children how to read?" That was the question Robby Bender had posed. I now understood it. I had discovered engineering. All I needed now was a technology of engineering to accompany my deeply ingrained urge to engineer. Linear programming and "branching" were methods, solutions seeking a problem. Good engineers don't buy solutions like that. But if you're going to be an engineer, you've got to get out of the laboratory, and preferably out of the university. It was hard to leave "true science" (and

Harvard), those warm havens where you're sheltered against the demands of clients. But engineering without clients doesn't exist. So I left, determined to develop the technology that was needed away from the distracting influences of science. As far as I was concerned by now, the Great Educational Revolution had not even begun. By 1961, I was saying:

"If you don't have a teaching machine, don't get one: don't buy one, don't borrow one, don't steal one. If you do have such a device, get rid of it. Don't throw it away, because someone else might use it."

## Back to Boston

*When they grow cotton in New England,*
*When Hell freezes colder than the Massachusetts Bay,*
*When the kudzu eats the ivy covering Harvard,*
*I'll be back, boys, I'll be back to stay.*
*I'll be back to stay, boys,*
*I'll be back to stay.*
*When the watermelon grows*
*Between the rose of Boston houses*
*I'll be back to stay, boys,*
*I'll be back to stay.*

*When they grow cotton in New England,*
*When Picket charges up Bunker Hill,*
*When okra grows in the yards of Cambridge*

 Human Incompetence

*And Jessamine flourishes on every sill,*
*I'll be back, boys, I'll be back, I will.*
*I'll be back to stay, boys,*
*I'll be back, I will.*

**Tom Gilbert**

But also, by 1961, I had developed what I thought (erroneously) was the true technology of education. You see, I didn't understand the difference between education as a whole and the tactics of development. I was trapped in one of those six blocks I've shown you before. But before we study it, suppose we digress a bit to see where we are (I think) going.

# CHAPTER THREE
## AN EARLY EDUCATION

### Heritage

*When climbing among the rocks and boulders*
*For whatever may be found,*
*It's hard to tell if you are standing on shoulders*
*Or merely on shaky ground.*

*Tom Gilbert*

### "Be careful not to make a show of your religion..."

My religious education began when I was about eight; my grandmother, Martha, started it all. She sent me to church and Sunday school, though she never went. I adored Martha Knight Barton and she me. Since my mother worked all the time, I spent much of my childhood in Martha's boarding

## Human Incompetence

house on Hampton Street in Columbia.

"I want you to go to church and learn what they have to say, but keep it to yourself. Don't ever, ever talk about your religion to anyone!" And, with one exception, she always enforced her stance.

The exception was certainly not my father's father. He was a religious nut who preached to anyone who might be near. But if he came to Martha's house, she would say, "Don't you open your mouth about God, Mr. Gilbert." More than once she threw him out.

My mother had taken Martha's training in these matters seriously, and I never heard her once discuss religion, not even on her deathbed. But my father was a card-carrying member of the Communist Party and an outspoken atheist. However, Martha adored her son-in-law, and when he pontificated about the foolishness of the Christian believers, she would hug his shoulders and feed him her best fried okra. "There, there, Frank. Don't get too upset."

I followed the teachings of my adorable grandmother throughout my childhood, I listened carefully in both the Presbyterian and Baptist churches where I went, and kept my mouth shut.

But I found the Bible fascinating, and secretly studied much of it, especially the New Testament. I even became a sort of amateur scholar, and at one time thought of this as a career—not that I would mention this to anyone. "Keep it to yourself, Tommy!"

Then in 1970 I ran across a new Bible, called the New English Bible (Oxford University Press). This was put together by a large committee of scholars who sought the earliest language available. They attested that this was the closest thing to the authentic Words. So I read it carefully. Here is how Matthew 6 reads in the NEB, the early part of the Sermon on the Mount:

"Be careful not to make a show of your religion before men; if you do, no award awaits you in your Father's house in heaven."

You will not find this in other bibles. Some skip it and others go on and on having Jesus pontificate about the hypocrites. Concordances say that Jesus was only refering to the Philistines. But the NEB makes it clear that nothing of the kind supports this argument. Jesus goes on to say that you must pray in private, give alms secretly, and call no man Father (rabbi)—but he does not say that this applies only to hypocrites. Martha could not have said it better.

## Don't you ever use that word again!

My grandfather, James Darling Barton, was a large cop who walked the beat for thirty years on Assembly Street, the poorest black neighborhood in Columbia. Once in a while, Uncle Jimmy, as everyone called him—except Martha, who called him Mr. Barton—took me with him. Uncle Jimmy had a terrible arrest record I was told, and I could soon see why. Once we encountered a fight in mid street, and Uncle Jimmy waded right in to break it up. Everyone knew him, and he soon had them all back in the shacks which they called home. And he went in with them and talked to them

Human Incompetence

til all was calm. I remember one woman saying to me, "Uncle Jimmy is the best man there ever was."

I recall once at about age nine using the word "nigger" in front of Uncle Jimmy as he mouthed his semi-illiterate way through the newspaper. "Wham." This huge, gentle man struck me in the middrift and I sailed across the room. "Don't you ever use that word again!"

Years later, I discovered that the reason Martha called Uncle Jimmy Mr. Barton was because he had made quite a few girlfriends down there on Assembly Street.

## The Land of Milk and Honey

My earliest secure memory was of my brother being born at home on Valentine's Day in Elizabeth City, North Carolina, as I cried on the cold, dark back porch until my father came home and let me in. I was four, and I never recalled Dick with much fondness.

*Tom's father (year unknown)*

But my father was a giant through much of my early years. He and all of his many brothers and sisters (twelve of them), and his father, the ever-preaching Mr. Gilbert, had all finished high school and were quite literate. The men all sold things as my grandfather labored away as a postmaster. Dad sold newspaper advertising and read the comic strips

Confessions of a Psychologist

Tom as a young boy, approximately five years old

on the radio on Sunday mornings. He also "wrote" a book—he called it *The After life*. It was a black book, hard-backed, with maybe 400 pages. Open it to any page and there was nothing but black paper. It was much more pleasant to remember him recounting over the neighbors' radios the adventures of *Dick Tracy* as Mom took us for a walk on Sunday mornings.

Mom had left school at the end of the eighth grade and would have followed the tradition of entering the mills to work had it not been for her oldest sister, Eve. Eve ran off with some guy and left behind a secretarial course, and Mom came to type over a hundred words a minute; she could even do this as she talked to you.

Every town down South had a cotton mill, and I learned many years later why that was. After reconstruction, Southern leaders made a bold decision; it wouldn't do for blacks and whites to work together in the fields picking cotton. So a mill was built in each town to employ whites to spin as the blacks picked the bolls. It was an economic disaster of course, but a social success of sorts.

At any rate, Mom worked as a typist until she met Dad and he took her away from all that. Pretty soon we were living in Winston Salem, N.C. until Dad got ambitious to move to the land of milk and honey. In 1932 we all piled into a pickup truck, two families and four children, I the oldest; and amidst all our belongings, we sailed off for California. Sister Eve had preceded us and taken up with a rich man and lived on Long Beach. We moved in with Eve and I started in an upper-class school. I adored it until the milk and honey showed up. Mom didn't get along with stuffy Eve and she had gotten us on welfare. At the end of the first school day, I received a quart of milk and a jar of honey. The other kids followed me home, taunting as we went, until I turned in anger and threw the milk and honey at them.

### Nostalgia

*My father burned his bridges behind him*
*With such alacrity and skill that*
*He was left stranded on a mound*
*In a sunless gorge, providing me (I thought)*
*A lesson. I cast my vision to high vistas.*

*More often now*
*The horizons I view lie beyond chasms*
*Untraceable over the spans of time I tore down. What*
*vision's left is needed*
*To leep the crevices that lace the edges.*
*So when the lights go off,*
*I hope they do it slowly,*
*Not quickly so I can't tell dim from bright.*
*If I had a choice between*
*The end coming on tip-toe or in a crash,*
*I'd take the stealth.*

Confessions of a Psychologist

*But for old-time's sake,*
*Burner Of All Bridges,*
*Come crashing in.*

*Tom Gilbert*

## Say, Brother, Can You Spare a Dime?

The milk and honey were not the last of my humiliations before we left the Golden State. Somehow, I spent a couple of days on the corner of Hollywood and Vine—my Aunt Eve later gave me the location—with a cigar box begging for dimes. If you can picture it, you would think I would have been quite successful. Age six, extremely skinny and freckled, with a great mop of red hair. But not many dimes were left. Things were really hard in those days.

I also remember her getting caught shoplifting. Naturally, my mother denied all these things when I asked her about them many years later. Or else she blamed them on my father's sister, Aunt Jean. But I'm certain that these experiences were enough to lead my mother to spend the rest of her life escaping the grasping threats of poverty. From then on, she worked every moment and pinched every dime.

But before she began the struggle, we witnessed the great earthquake of March 1933, and then crawled back into a pickup and headed for South Carolina, the land of grits and okra, where Martha had a boarding house and Uncle Jimmy was a respectable cop. Broke, we moved in with Aunt Net, who also had a boarding house down on lower Main Street. It was called the Dixie Hotel, five stories tall and no elevator. The most I remember about the Dixie, besides the huge rats

 Human Incompetence

which scampered across the floors, was the sight of Franklin Roosevelt as he made his way up Main Street smiling and brandishing his cigarette. Down on the first floor lived Mr. Patton, the General's brother, who despised children.

### Road to Leisure

*At the crossroads of Debility*
*Stands a sign for all to read*
*(An antipode to an older creed):*
*"To each according to his ability,*
*From each according to his need."*

**Tom Gilbert**

# CHAPTER FOUR
## KID NEWTON AND THE "MIND-BODY" PROBLEM

I had a terrible time with physics and it took me many years to understand why: I had gone about it 180 degrees in the wrong direction. But so that you can better understand what I mean, and this may be my most important lesson for you, I'll first tell you about the contributions of an extraordinary person. His name was Isaac Newton.

Recounted briefly, his early life is almost too much to be believed. In 1666 when he was about twenty-one, this underclassed kid (his mother remarried a man of substance) was sent home from college to escape the plague. Two years later, working at home all by himself, with virtually no books, he completed the science of physics, perfected calculus, and created the science of color and optics. And I mean completed. He dabbled with these things a bit from

 Human Incompetence

time to time, but after that he wrote about religion and the mystical. He lived into his eighties and was honored far and wide, but his work was virtually complete at age twenty-three. How's that for genius? He responded to the Archbishop of Canterbury's inquiry about what had driven him by saying, "I wanted to make the Universe safe for God." And he had done it all in a slim volume called *The Mathematical Principles of Natural Philosophy*. And don't think that modern physics uprooted him— by Einstein's own admission, he merely perfected what Newton had started.

I have spent much of my career trying to understand how exemplary performers work. But I ignored Newton because it seemed that he was way beyond "exemplary." Exemplary means that you can make an example of them, and this is hard to do with certain top performers. Try to emulate Babe Ruth, but Brooks Robinson is truly exemplary and you can greatly improve your play at third base by emulating him.

Why did I have so much trouble with physics? I have been a superb student of most things, but this subject stumped me completely. But now it comes to me. Recently I took up Newton's book to see what was so hard. What an astonishment; I can still hardly believe it.

Discount a couple of things. The professors worked so hard to make physics difficult. Almost all of them do that, but I had early learned to escape that ruse. The books are written terribly, but I had become a good reader. One trick that helped me was Jim Simmon's "translate it into plain English."

But I was not prepared for Kid Newton's simplicity. He

starts out with his famous three laws of motion. Ready yourself for the first one, slightly reworded to impress you even more by its utter simplicity: "A thing just sits there until you push it."

Okay, you don't think it can be that simple? Here's the whole thing: "Every body continues in its state of rest, or of uniform motion in a right line, unless it is compelled to change that state by forces impressed upon it." Need I go on?

Here's the second law; "The harder you push it, the more it moves."

Try the third law (you should be in the swing of the thing by now). "Whatever pushes it is slowed down." It is said that the Londoners of 1776 spoke with a Southern accent. Could this have dated back to Newton's time? Can you imagine him saying, "A thing jes sets thar til ya'll pushes hit."

Thank God for the plague. If Kid Newton had stayed around in college for a couple of more years he may not have been able to make things so simple.

But I was terrible at physics; you see, I was a natural-born psychologist. I came out of the womb watching animals and people behave. They don't follow these laws very well at all. Certainly a cat will fall in a vacuum as fast as a feather or a lead ball (I'm told), but there the communality quickly ends. For one thing, it lands on its feet. Indeed, for cats we could just about reverse Kid Newton's laws.

 Human Incompetence

"A cat will keep changing its motion until you impede it." And on and on: "If you push it, it will push back."
When you drop it into a Skinner box, the first thing it does, after a little sniffing, is move about. If it didn't, you could never train it.

But let's not worry ourselves yet about the differences between Newton and Skinner. The similarities are what begin to impress us. Each followed three rules that the old philosophers of science said were critical to any good scientific theory.
- The first of these rules is called *parsimony*, which means "stingy." Do not use two ideas to explain something when one will do.
- The second rule is called *elegance*. Don't fit your ideas together to make an eclectic mess. Rather, make sure they fit together in a coherent whole. That's pretty simple, and Kid Newton (and Skinner) was nothing if not coherent.
- The third rule makes sense. It's called *utility*. Don't posit anything that's not useful to your purpose of advancing the theory or applying it. They tell me that there couldn't even be television without Kid Newton's science, so I suppose it had utility.

One of my favorite poets is Edna St. Vincent Millay. Some stupid critic wrote rather snobbishly that she had written only a few great poems. Who else has? Ah, look on Kid Newton's slender volume, and Euclid's. Millay wrote:

*. . . Oh blinding hour! O holy, terrible day, When first the shaft into his vision shone.*
*"Of light anatomized! Euclid alone has looked on Beauty bare. Fortunate they Who, though once only, and then but far away,*

# Confessions of a Psychologist

Tom in front of Christmas Tree when a young man

*Have heard her massive sandal set on stone."*

Again such parsimony and elegance: Two points form a straight line. Three a triangle. Maybe we could write, "Newton alone has looked on motion bare." Or, "Skinner alone has looked on Psyche bare." So that was all that physics amounted to? And I began to understand it. And why I had such difficulty. And don't believe that Einstein made things much more difficult. He just tightened up a few things and $E=mc^2$ is about as simple as they come once you understand it.

## A Visit to Kid Newton

It's not that I got the wrong answers in physics, but that I got answers that were the exact opposite of the correct ones. Why had I such difficulty? I suppose it was the way I came out of the womb. I have found that people seem to continue the way they begin, though many psychologists would snort about my evidence as being anecdotal. However, with seven children and the pleasure of observing many more, that begins to amount to a lot of anecdotes.

I gave Adam a dime-store horn when he was three, and now he is considered among the top recorder players in the

 Human Incompetence

world. Little Rob came out drawing pictures, and he continues today as a successful illustrator. Sarah was telling stories from year one and has just finished her fourth novel. Micah and I were writing lyrics together when he was very young, and he just won a national prize for lyrics. It was obvious from the outset that Marilyn's son Billy—twice on the cover of *PC* magazine—would have invented the computer if there hadn't already been one. And her daughter Andrea, now a successful lawyer, was heard at five in a sandbox making fine distinctions as she told another little girl, "I'm not a lesbian; I'm Jewish."

I think I enjoy the memory best of my oldest daughter Kathy, who is now South Carolina's administrator of remedial reading. She sat on my lap and seemed to read perfectly at age six until the teacher called and said Kathy couldn't read. So then I looked carefully. She pronounced the words perfectly as she squirmed and stared out the window. Obviously, she had memorized the lessons.

I came out of the womb injuring myself. Indeed, I could well lead the world as the most accident-prone person—I just spent two summers in the hospital recovering from fractures. There are few bones I have not broken. The first law of motion is only now beginning to dawn on me. I have lived in a malleable world for lo these many years. When you approach obstacles, they just move out of the way, don't they? Cats do.

So I was really primed in my head to look into the "mind-body" problem, whatever that is. All those things just sit there until you push them, do they? Well, not the ones I focused on. The apple may have fallen on Kid Newton's head but it would have sailed purposefully into my arms.

If all those things just sit there waiting to be pushed, what are all those things getting up to push them? Ah, then I began to see it. It is how we measure things and know things. Let's take one of those solid, unmoving objects that Kid Newton says will just sit there. What do we know about it in a general sort of way? It just sits there—but for a while. It endures in time, or else we would not know it. Time (T) becomes the first fundamental property we assign to it. But the *mind*, the thing looking at it, doesn't endure in time and doesn't even seem to. Time occupies it.

But how do we measure time, we up-and-at-em types? We jiggle; we jiggle our fingers, our tongues, use clocks to jitter for us—it really doesn't matter what, but we jiggle: "It's still there, it's still there, it's . . ." Not much more to it than that. We jitter. Some things that aren't animals jitter also, but all animals must. This is fundamental. Pigeons, sea urchins, and people jitter. To have an agreeable system for measuring variances in time in the physical world, we must fix a system of jitter.

Experiments have shown that if we stop the tiny vibrations of our eyes we can no longer see. But while we fix the system of jitter we use for telling time, we can vary the jitter as we will. So while time is variable in the fixed system, it is fixed in the "mental" system. But in order to measure changes in jitter, we must fix our times. "How frequently does the animal jitter per minute?" we ask. Interestingly, the notation we use for jitter—for frequency—is just the inverse of time, or I/T. Let's call this fundamental property of mind, or of

behavior if you will, "tempo," so as not to confuse it with physical matter. Why "tempo?" It only sounds a bit more sophisticated than "jitter," but we might let "J" stand for it. Can we take this as a clue—this inverse property of time and frequency? Are the fundamental properties of the mind just the inverse of the fundamentals of matter? It has appeal. Will it hold up?

Let's take another seemingly fundamental by which we know behavior. It doesn't just sit there until you push it, but it "gets up and goes." Indeed, if you push the cat, it's likely to push back. But pushing something of the physical sort is how we know another of its fundamental properties—it has mass, or M in the physical notation. A rock only accelerates (or decelerates) when some outside force acts upon it. But the cat exerts the force all on its own. In the physical system, force has the inverse property to mass. Let the cat stay the same, in a fixed period of time, it nevertheless exerts a force. Since we needn't (and shouldn't) use the notations of physics for force and acceleration, why not call this property of the animal, "animation." When something just gets up and goes is when we know that it is not merely matter but has a mind. And we will let "A" stand for this dimension.

Another fundamental property of matter is its extension in space— length or distance which the physicists notate as "L." To measure this property, we need to fix our stance. Indeed, the mind is built to do this. A pack of cigarettes one hundred centimeters long remains to appear this length even as we move away from it. This "size constancy" seems to be a fundamental property of the mind. Now, emboldened by our efforts above, let's consider what the physical inverse of length is. *Curvature* (I/L), it's called in physics, certainly not a fundamental in that science. But for us to estimate physical length, we must fix this fundamental of the mind. Curvature is a change in direction. And certainly, this capacity to change direction is fundamental to the mind, to the behavior of animals. Push the ball and it will roll in a steady direction until it encounters some other force. Push the cat, and there is no telling in what direction it will move and it may even push back.

The most remarkable characteristic of the mind are the changes in the way it looks at things—in its vantage points—its continuing "curvature." This dimension is as complex as the dimension of length is simple. To designate a good word for this fundamental property of the mind, let's call it "vantage."

So here are the fundamental dimensions of animal behavior, or of the mind if you prefer.
- Tempo, with one convenient property being the probability of a response.
- Animation, with one convenient measure being the acceleration of a response.
- Vantage, with convenient units to be established with any amount of agreement.

One set of properties might be the levels of vantage described in Gilbert's *Human Competence*, pp 114.

MORE TO COME ... [SIC]

 Human Incompetence

### *Vantage Points*

*The eye can travel to a star*
*So fast that Science cannot measure;*
*How vast a scope the eye can grasp—*
*Infinitudes of pounding pleasure.*

*And Science can little comprehend*
*What tiny things the eye can see:*
*Imaginary rebuffs of love—*
*Infinitesimal misery.*

**Tom Gilbert**

# CHAPTER FIVE

## EIGHTY-EIGHT YARDS OF ENGLISH GRAMMAR

In 1958, while serving B.F. Skinner as a fellow at Harvard, I had to make a decision to stay on in academia or go out into the so-called real world. I hedged my bets and went for a rather heroic effort that would allow me to stay within a university while still devoting my work to making the results of my efforts more likely to be useful out there in the real world. Heavily influenced by my five-months' summer stay at Bell Telephone Laboratories on my way to Harvard, I devised a plan for a behavioral research laboratory that would encourage the flow of pure research through several stages until its products became usefully applied. The one school that seemed most equipped to support this effort was the University of Alabama, especially with the encouragement of Paul Seigel, the head of the Department of Psychology, and John M. McKee, the head of the Alabama

 Human Incompetence

Department of Mental Health. Paul arranged a one-year faculty appointment for me while I wrote the proposal to the National Institute for Mental Hygiene. The proposal was quite long, and it detailed exciting possibilities for research in both education and mental health.

Just as one example, I had convinced Charles Slack to join me. Charley had been doing some interesting things at Harvard with juvenile delinquents—not ordinary delinquents, but fifty young men deemed incurable by the prestigious Judge Baker Guidance Center in Cambridge. Charley didn't believe in halfway measures. While in Cambridge he managed to get a research grant for his bad boys and sent various people, such as ministers and social workers, out onto the street to engage the boys. As they walked into Charley's offices, they were immediately handed money to reinforce their attendance. They were then sat at a screen that asked them to talk about themselves. A counselor sat behind the screen and delivered money as the bad lads talked. Charley had reinforced them for attending a therapy session and for talking while they were there. He delivered the money (nickels—this was in the 1950s) on a variable ratio schedule, meaning that the delivery was randomly contingent upon the rate of response. This is the same schedule of payoffs used in the gambling halls of Atlantic City.

Later, as he had established regular attendance, he had the boys join his Center's corporation. Down one side of the letterhead were the names of the boys; down the other side were the names of the board members—names like Eleanor Roosevelt, Admiral Hymen Rickover, B.F. Skinner, and on and on—a veritable *Who's Who*.

Charley saw to it that the boys did many things in helping the community (for which they got paid) and even saw to it that they all passed their driver's license tests. Charley's criterion of success was not a single repeat offense among his fifty boys for a year. Alas, one kid lost his driver's license and got caught using someone else's, not knowing that you can lose your license without losing your privilege to drive. The kid's offense went against Charley's experiment. The National Institutes of Health, looking for an excuse not to fund Charley again, held him to his own rigorous criterion and refused to continue support for the second year. Just think of it; he prevented forty-nine *incurable* delinquents (so judged by the prestigious Judge Baker Center) from repeating an offense, but lost because one kid used someone else's driver's license (he hadn't stolen a car as these guys had been wont to do).

Scratching for funds here and there, Charley kept to his standards for another year (this time, fifty successes) before he departed for Alabama where he too received a temporary appointment and helped with my proposal.

The proposal went off in time, asking for $3.5 million (remember, this was 1959!) and with a board of directors, which was a *Who's Who* without exception. Participating in the proposal was every social center in Alabama, such as the Department of Mental Health and the Draper Correctional Center (prison for first offenders). The heads of all these agencies were well informed and wrote enthusiastic intentions of support. The university's President and many other officials gave it similar support.

 Human Incompetence

The National Institute sent forth a committee of nine, as I recall, who spent a lot of time talking to everyone. Poor Paul Seigel, the head of the Psychology Department, barely had time to attend to his official duties, and even his wife Helen, came to hate me. The only really good thing to come out of all this, I suppose, is that both Paul and Helen became fast friends of mine after the Center proposal was all done.

Rumor had it that the committee was going to shoot us down, but Charley Slack did what he could to prevent this by sailing off to gather the commitment of four more *Who's-Whoers* to form an oversight committee. If they deemed at the end of each year that we were not on target, the Institute would be under no further obligation to support us.

Finally, the committee voted for us six to three, but the minority wrote a strong negative report that the Institute used to kill the project. And so there I was, a complete failure and stuck with a hard question—what to do with the rest of my career. I was determined not to return to academics, so I rented an old house in Tuscaloosa and set up my first business: Educational Design of Alabama. And before long I was making a living at it. Programmed instruction had caught on, and I was the leading light in that business by 1959.

Some of the agencies in Alabama used me as a consultant, especially the state prison for first offenders, Draper Correctional Center, under that enlightened warden, John Watkins, and the State Department of Mental Hygiene under John McKee, an old friend who had gotten his Ph.D. with me at the University of Tennessee. We set up a *programmed instruction* lab at Draper Prison and fixed it so that the boys could escape field work if they made progress with our school materials. And they even began to program materials of their own.

At some convention, an entrepreneur, a great box manufacturer, anxious to make millions in programmed instruction and with his cardboard teaching machine, asked around for price quotes on lessons in English grammar. These turned out to be far too high; you would have thought he was asking for nuggets of gold. I told him about the Draper school and how I could get the inmates to do it for him much cheaper. He almost settled on a deal for about 3000 frames of grammar lessons for $500 when the Draper boys were disturbed by what he meant by a *frame*. (In programmed instruction, a *frame* is an area on the paper presenting information or a question with a blank to be filled in by the student. It also contains the answer to the previous frame). "Are these three-inch frames, two-inch frames, or one-and-a-half-inch frames?" The entrepreneur settled for the one-and-a-half inches (these being rather typical), and the Draper boys rewrote the contract to read $500 for 88 yards of English grammar. And he actually signed it! He complained to me at first—"what will people say?" I reminded him that no one need see the contract, only the 88 yards.

Someone pointed out that 3000 yards is not exactly 88 frames, but just think. These young first offenders lacked much education and couldn't be expected to be dead accurate with arithmetic.

So what a way to start an educational revolution—by the

 Human Incompetence

yard! Maybe they could have gone all the way if the contract had been for a hundred yards!

And don't think that their programmed instruction efforts were poor. One kid wrote a piece on how to blow the trumpet. To indicate how one should form the embouchure (the lip formation), he asked the students to imagine that they had some chewing tobacco on their lips. To form the embouchure, "Just imagine you were spitting the tobacco off your lips." (Find a better way to describe it!)

I think my most vivid memory of this Great Educational Revolution was the sight of Charley Slack displaying the box manufacturer's teaching machine at a convention booth I had rented. It was a cardboard box with the required window and it opened up so that you could insert a roll of programmed instruction in it. Charley would hold it up and point out to people that it would not only hold the teaching materials but two sandwiches for lunch as well. Then six-foot-six Charley would dance up and down on it and cry, "Come and get your indestructible teaching machine!" I don't think we sold any. Alas, that may have been the beginning of the end of the Great Educational Revolution.

What was wrong with the Great Educational Revolution— teaching machines and programmed instruction? It caught on at first and promoted great fanfare, so why did it die?

First of all, I think, it promoted the wrong things. The teaching machine was certainly not required. Dumb as everyone else, I had a machine of every variety. One was quite large and heavy and later, after I had abandoned the machines, Bell Labs insisted on buying my large one (for

$6000!) and they tried to reproduce it, but theirs never worked. I abandoned mine in stages. First, it was so cumbersome and expensive that I could afford only one of them, so I resorted to using a file folder with a window cut into it. The students would simply slide their programs forward until the current frame appeared in the window. But one bright student tossed her folder away and simply took the materials one frame at a time, leaving all the others exposed. That worked as well as the machine, so I was thrown back to pre-machine conditions.

Thus, the only remaining challenge was how to program the materials. And then it soon dawned on my simple mind that not even the frames were needed if the teaching materials were properly constructed. Alas, for me, The Great Educational Revolution was over and I was back where we started from—how to construct good teaching materials, and that was sufficient challenge that I didn't need to be encumbered by the frame format.

The dumbness of the whole thing was, I think, well summarized by a verse my friend Jim Evans wrote:

> *B.F. Skinner left a hole*
> *In a sentence one day,*
> *And a thousand disciples abandoned*
> *Their jobs to keep it that way*

So I began to rethink the whole issue of how we should best design teaching materials, and I came up with a system I called *mathetics*, from the Greek meaning "having to do with learning." I even published a *Journal of Mathetics*. I got

 Human Incompetence

two issues out using money from a program I sold in basic algebra. But someone sued me claiming mathetics as belonging to them. I won the suit but it left me too broke to continue publication. Alas.

Mathetics had nothing to do with presentation formats. It simply followed some basic rules we had gained from the learning laboratories. For example, one rule was that in teaching difficult stimulus discriminations, you begin with the hard things first. Thus, in teaching the multiplication table, you begin with the 8's, 7's and 6's (9's and 5's aren't difficult because there are memory aids for them.) And you teach the pairs together in groups—you've got to learn to tell the difference sooner or later, so the sooner the better. And so on and on.

Certain chains of behavior you teach backwards—meaning you begin at the end and gradually work your way backwards through the chain. Thus, to teach kids how to tie their shoelaces, you tie them to the last step and have them complete that step—pull the bights to tighten the laces. Then you go to the next to the last step, and so on. Chaining doesn't apply very often because we have language that will aid the process. Indeed, about the only places I have found that backward chaining really works besides shoelace tying is in teaching dentists to mold dental braces and musicians to play certain sequences.

Since the *Journal of Mathetics* appeared, I have identified some fifty principles that apply to the design of instructional materials, though only six or seven apply to any given problem. If used properly, these rules will allow you to conform to *Gilbert's Law of Training*, which says, "The very best training is always the shortest training." By following these rules, I have been able to teach the composition of seventeenth century counterpoint and harmony to the standards of Bach in about ten minutes; tossing a wad of paper unerringly into a basket across the room in about five minutes; cooking a very fine French omelet in a few seconds; greatly improving skills as a third baseman in minutes; learning the anatomy of the cardiovascular system in about thirty minutes; and teaching the multiplication tables to fourth graders in about an hour with no forgetting. And on and on; this only begins a very long list. And none of these achievements requires any preconceptions about the formats of presentation.

You would think that mathetics would be very easy to sell, wouldn't you? But, alas, I haven't found it so. In the early days of programmed instruction, some entrepreneur offered seven or eight people $1500 to design a lesson to teach the color code for electrical resisters—he wanted to see who could do it best. My effort took up only one page, and in five minutes you could master the code with no forgetting. One program required four hundred frames, and forgetting was still a problem. The entrepreneur got very angry with me for shortchanging him—only one page! The four hundred-frame guy won a special reward. Alas.

To test my method, try the lesson following (the color code applies to the strength of electrical resisters—color bands stripe the resistors and stand for numbers). Here you go:

 Human Incompetence

> One ***Brown*** *Penny*
> A **five** dollar bill is **green**
> An **eighty** year old man has **gray** hair
> **Zero black** nothingness
> **Three oranges**
> A **blue** tailed fly has **six** legs
> A **yellow** cab has **four** wheels
> **Seven purple** seas
> A **white** cat has **nine** lives
> A **red** heart has **two** parts

Read through this list a couple of times and you won't forget it— especially if you have occasion to use the color code. There is another memory aid that is incorrect: "Black bruins raid our yellow grain, blue violets grow wild." The fourth word is *orange*, the fifth word is *yellow*, and so on. When I tried my method out with secretaries at Bell Labs, they were faster than old-time electricians who used the black-bruins aid—you see, these guys had to go through the sequence to find their aid, the secretaries didn't.

The program was eventually bought by Bell and Howell, and I'm told that some one hundred thousand students used it there. That was some success, at least.

I have found about seven memory aids that are useful, but you must use the correct ones if you are to be successful. The popular ones that memory artists rely on aren't the most useful ones. But I will tell you a story that illustrates the attitude of many psychologists specializing in learning have about memory aids.

I was at a meeting of psychologists during the early days of programmed instruction. This meeting took place at the University of Pittsburg and consisted of high-level professors who had made a name for themselves in learning theory. Two graduate students were allowed to attend—one was Jim Evans, the author of the verse I show below. I had told Jim about a colleague, a professor down at the University of Alabama, name of Oliver Lacey. Oliver would have a few too many drinks and pronounce himself a solipsist. *Solipsism* is a belief that all reality exists only in one's own mind and nowhere else. Somehow, this appealed to Evans who was getting into his cups.

One of the learning theorists there had just published a book on learning, and in our discussion I asked him why he did not treat memory aids in his writings. The professor became very haughty and denounced memory aids heartily. He said that they were a form of cheating and poked me to bring home his point. Evans countered with a threat:

"Olimer's going to forgit you!" he said with blurred emphasis.

Maybe Oliver did forget him. I can't remember who the guy was.

### Gone is the Gilbert

*Gone is the mile of measured length*
*And pint and pound and peck;*
*Gone the gilbert of magnetic strength,*
*The dram no more correct.*

*And though the inchworm does not change*
*As the inch is banished,*

 Human Incompetence

*Won't the world seem wondrous strange*
*Now the foot has vanished?*

*The meter rules where feet once tread,*
*Enthroned beside the liter.*
*Oh strange, indeed, for rhyme has said*
*The foot will size the meter.*

*Tom Gilbert*

## Look Before You Listen

But Oliver Lacey helped me confirm an idea that was slowly shaping in my slow head. The interview is the most important tool most people use to learn how experts do their jobs. I had even begun to write a book on how to interview. But one of my employees, Betty Hatch, accompanied me on an interview with a power company executive. As he answered my well-formulated questions, I kept interrupting and asking him to go show me what he was talking about. Later, Betty shook her head and commented about what a terrible listener I was.

"Tom, you are a visual animal; you don't listen well at all!"

Indeed, almost everything the executive told me didn't match well with what I later saw. It slowly dawned on me—the real method of science is observation, not interviewing. You simply can't rely on what people tell you.

So Oliver helped me put this idea to test. He had once been the seventeenth seeded tennis player and had managed to get a professional tournament to come to Tuscaloosa. As we sat in the bar, I asked him, "If we interviewed these top pros, could they tell us accurately how they moved their feet and arms as they played?" "Let's test it," he challenged. So I drew up a set of questions and we called in the pros one at a time and interviewed them, I asking the questions.

Later, we observed them and conducted a correlation coefficient between what we saw and what they actually did. But there was a small negative correlation—these guys just didn't have it right.

But when you think about it, why should they? With five thousand fans at Wimbelton, who are the only people not able to watch their feet? The pros, naturally. If they were watching themselves, they could never compete.

A client of mine at a large manufacturing company put my observation theory to test. The company had a factory in which the machines just weren't performing well and they called in their brilliant troubleshooter to find out why. Before she entered the manufacturing room, they interviewed her for two hours, taking the whole interview on film. "What did she expect to find? On and on she went (they later edited the film to thirty minutes).

Then they opened the door and let her enter the manufacturing room. Just as she walked in, she gasped and pointed to something, "There is the problem, no question about it!"

It seems that on each machine there was a tube of fluid. "It's blue," she said, "and it should be purple!" It has too much sulphur in it." (I'm not sure of the colors or the sulphur, but she saw it instantly.)

 Human Incompetence

Why hadn't she mentioned this in the interview? "I have never seen it before!"

Since those times, observation has been our principal tool for identifying what exemplary performers do that set them apart. Three things we have found.

First, exemplary performers love to be observed, though they rarely are. They may be a little uneasy at first, but within thirty minutes they are relaxed and talkative. But be careful not to listen unless you are also observing.

Next, almost every important thing you see you will never uncover in an interview. Why? Well, just to begin a long list, it never occurs to them to describe their *secrets*. And not because they are trying to hide them. "Doesn't everyone know that?" is a common response.

Finally, the critical things you find in observing the exemplars are very simple. Exemplars do their jobs more elegantly—much more simply than the others. They don't make life hard for themselves.

We have found only two performers unwilling to be observed. They freely submitted to interviews, but not to observation. These were the two best-selling Yellow Page advertising reps we ever knew. But later, both landed in jail because they were making kickback deals with customers. These were not exemplary performers—you wouldn't make an example of them.

Alas, managers rarely observe these exemplars; they travel with the poor performers instead. What they could learn if they only reversed their routines. I'll just give you a few examples.

## The Pharmaceutical Reps

Pharmaceutical companies employ sales reps to visit physicians. In one company, they kept careful records of the prescriptions in each rep's area. The usual reps—we found as we observed them—spent a lot of time talking to doctors whom they were comfortable with. But watch the rep with the best record in his area.

He would walk into an office—usually of a doctor who wasn't using his drugs—and give literature and samples to the up-front assistant and explain it to her, asking her to be sure to convey this to the doctor; then he would be quickly off to another doctor's office—but not before leaving a gift behind for the assistant. This way, he covered far greater territory than the other reps, and he would follow up with telephone calls. He would also nudge non-users to try his stuff once or twice. And he would also go back to see a doctor if there was any sign of interest. He rarely saw a doctor who was a committed prescriber.

## The Candy Salesman

A candy *salesman,* as his company called him, did not conform to his job description, which said that his job was to sell candy. "I don't sell candy and gum; the stores buy my Dentine whether I want them to or not. Watch me." And we did. He would enter a store, find a place to display his goods so that they were at eye level, and then he would break up the packs of say, gum, and rearrange them in a

 Human Incompetence

strange order so that the pieces stood out visually. Then he would hang little handmade signs on them which said such things as, "*This is the best gum there is!*" Sometimes the store manager would clean up after him, but he would have attracted quite a few customers before this happened. And if his supervisor was planning to visit the store—which was rarely (remember, they don't visit much with top performers)—he would clean up the day before so that the boss would never see what he had done. "The displays don't have to be attractive," he would say. "Just so as they attract attention."

## The Forklift Truck Sale Reps

The three best forklift truck sales reps we ever observed confirm something we have learned well. The behavior of exemplars does not set them apart—it's what they accomplish. Managers will tell you, if you ask what the top performers have in common, "Nothing; they are all totally different." And in a way they are right. But watch with me.

The first rep, who was considered the best in the world, would be driving at high speeds down the snowy backroads of New Jersey and come to a screeching halt. "Look at that old abandoned warehouse," he would shout. "They are about to reopen it!" All he saw were some footprints in the snow entering the warehouse. In a day or two, he had sold two forklift trucks.

A second top rep would spend the morning going through all manner of paper looking for signs of new businesses opening (the first rep hated such studiousness) and he would take careful notes. Then, about noon, he would begin to call on his list of new prospects—he rarely saw anyone who wasn't a hot one.

The third rep spent his mornings on the phone, but doing the same thing, looking for new prospects. Then off he would go. He loved the phone and was quite good at it. The first rep said he hated phones.

And the average reps? They would just call around rather randomly. "How are you fixed for your forklifts?" This kind of random behavior sometimes pays off—like drawing to an inside straight. Idiots do it.

## The Four-Foot-Ten Jewel

Now let's watch Jewel sell Yellow Page advertising space. She sells about three times as much growth advertising as the average rep. How does she do it?

First you need to look at Jewel and wonder what kind of impression she makes. She is black, about 4'10," and dresses rather dumpily. But that only begins the list of her appearance problems. She also has a lisp. Also, as she enters a customer's office she doesn't shake the offered hands. So how does she do it? Indeed, how did she get hired?

She was a telephone sales rep (her lisp wasn't quite that bad) and would have remained there had a lawsuit not demanded that the company employ blacks and women. The personnel director delighted in hiring her—she was a token black and a token woman, all wrapped up into one, who was bound to fail. So how did she succeed so well?

Human Incompetence

You would have to observe her—no amount of interviewing would have helped.

Let's watch her as she enters the office of the president of a very fancy realtor. He is naturally taken aback by her appearance, but not for long. "Mr. Jones," she said, "all your advertising is wrong. You are trying to reach ordinary buyers, but you really want to reach people with expensive homes to sell. That's your business—selling top-of-the-line properties." She has his attention and he is no longer aware of her appearance.

Within an hour she and the realtor have designed a completely new advertising program with a lot of growth in it. This she had already sketched out for him.

"My God, she was great. I would like to hire her!" Not likely. She was earning $150,000 a year, with her high commissions.

And Jewel did one more thing that helped her a lot. She employed some high school kids to call on accounts that she was pretty sure she could renew. They did that quite well. It most probably would have been against company policy had the company been smart enough to think that someone might do such a thing. She also hired her own clerk to do her paperwork. The company certainly would not have approved of that either. But, you see, the top reps had much more paperwork than the average reps—why? Because they sold so much more.

Now let's see what Jewel accomplished. First, she made an appointment with the top decision maker—few of the reps bothered to do that. They would just enter a business and hope to hit it lucky. Drawing to inside straights, again. And next, she had studied the business and found out what the customer really needed—she hit his hot button. Some reps would have gone in and suggested that he buy some new space—"What do I need that for?" And finally, she was all prepared with the details of a plan. Many reps would have waited until a sale was already made. Don't look at Jewel's behavior; observe her accomplishments.

And look before you listen!

## Ohm, Sweet Ohm

*What of the watt as the horsepower's source,*
*With kilos and millis, the unit of power!*
*Behold the bold newton as a measure of force,*
*And the duel of the joule and the killowatt-hour!*

*Hail pascal, that rascal, that prince of a measure!*
*Frown on the pound and the passing pound-force.*
*Let a jolt from the volt spoil faraday's pleasure,*
*And pamper the amp that lights up the course.*

**Tom Gilbert**

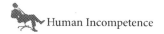 Human Incompetence

## Show Me Now and Teach Me Later

Shortly after I went into business in Tuscaloosa, I employed a young girl as an assistant, S.K. (Katie to me) Dunn. She was generally unemployable, but she was a cousin of a friend, and one interview convinced me that she was at least bright. As it turned out, she was brilliant in a strange sort of way. She had little formal education, although she came from an upper class family. An uncle had been the ambassador to England and other family members were also distinguished. But, somehow, she had almost completely escaped an education. She could read well and little more than that.

I remember, at her request, giving her some psychological tests. Her scores were so unique they could have made a publishable book. One test she wanted to take was a strange diagnostic tool called the Szondi Test. You won't believe this one. The Szondi contains photos of people with every variety of mental illness you can imagine. The subjects are asked to identify the illness of the person in each photo. Any they get right are supposed to be indicative of their own psychological disorders—or at least their proclivities. She had an astounding score; I'm sure no one ever did as well. She got them all right! I wrote to the Szondi people and they responded by saying that this was impossible. The odds were certainly well against it. When I told her about her results, she clapped her hands: "Oh! I passed my Szondi test!"

On another occasion, Katie was at a party with a number of us psychology Ph.D.s and she asked four of us what kind of research we did. We each carefully described our experimental work. I had been studying the cycles of eating behavior in rats. Paul Seigel was into "micturation," I think he called it—urination to you illiterates. Another person was investigating certain sexual habits. The fourth was studying defecation.

"Ah!" Katie said with great appreciation. "You guys are studying the basics: eat'n', shit'n', drink'n', and fuck'n'!"

With all of our emphasis on training in the days of programmed instruction, Katie had some influence in turning my head around and taking job aids more seriously. When she would ask about something in the office, I would usually launch into one of my professorial lectures, and this seemed to annoy her a bit. Finally, when she had asked about how to do something of importance and I went into my usual pontification, Katie said with bridled impatience, "Show me now and teach me later!" That should be engraved above the doors of every Training Department in America.

But Educational Design grew and was finally bought out and became TOR Education and we moved to New York and Connecticut. Katie and I became lovers, and she bore me one of the sweetest children ever known to humanity. Jessie loved art and she loved order. For years now, Jessie has been managing a very fancy art gallery in San Francisco, one that specializes in early nineteenth century landscapes. If you want to buy just a small sample, take along at least $50,000. Katie has shown some proclivity as an actress and playwright and has appeared off Broadway on several occasions.

MORE TO COME ... [SIC]

# CHAPTER SIX

## *MOM*—OR METHODS OF MANAGEMENT

**How to Manage the Cleaning**

My mother went into business for herself in 1939. First she built a small tourist court—we had moved into the country out on Highway One north of Columbia and had a spot of land, not far from Fort Jackson, which was quickly filling up as one of the large army bases. And she also received a commission to operate a tailoring and dry-cleaning shop on the base. This quickly grew to seven or eight shops and she made quite a bit of money doing this. She was allowed to expand simply because she operated the most efficient shops at the Fort. It was here that I received my first work experience and plenty of it.

I did everything, leaving school to come straight to work.

# Confessions of a Psychologist

*Tom's mother (year unknown)*

I operated sewing machines, used the presser, and took in clothes, you name it. And we worked late. Sometimes the line of soldiers stretched almost out of sight, leaving the heart numb.

But never once did mom touch a thing. Many years later, while she was visiting us, she asked to use our sewing machine and I had to teach her how it worked. Mom managed and that was that. She always seemed to be an efficient manager. In all those years she had practically no turnover and her employees adored her. She paid them well and they even came home with her on occasion—they were black, and remember, this was South Carolina in the forties. At the time, I thought how much she would have gained had she worked in a large company where they had really professional managers; little did I know.

What did she do that made her stand out as a manager? What did she do that practically no manager does in those large companies I have come to know so well?

Okay, she was nice to people, but that wasn't what made her so good. What she did was to perfect what I call the Three I's. She gave everyone excellent Information about what was expected of them and how well they performed. She provided them Incentives for working well; there was always extra pay for those who turned out a lot of work. And everyone received excellent Instruction. One of her

 Human Incompetence

ladies spent a good bit of her time teaching the others how to perform to top quality and speed. The "Three I's"; almost all you need to manage well. And the three things that you find so little of in those large gleaming buildings of work. Let's take them one at a time.

*Tom's mother, March of 1968*

We mistakenly call this the Information Age. It is no such thing. We so easily confuse information with data. Hidden in those huge stacks of data the so-called Management Information-Departments give us are tiny bits of gleaming information that management rarely sees. This is the Data Age, not the Information Age. Information is what gets through to us. The more data they pile on managers' desks, the less information they have. It makes them feel good, but it doesn't tell them what they need to know. They call it the "information superhighway." It's the data back road, a dumpy thing made of dirt, and we surely need to learn the difference.

I have been in almost every kind of business there is over the past thirty years, and rarely have I known people who really knew what was expected of them or who had the information (not the data) they needed to do their jobs well. Nor do they often know how well they are performing. I've written a little verse that says it all:

## Haystacks

*Data, like the hay, is usually dry*
*And piled in stacks and measured by the bit.*
*But how like the needle information is. It*
*Always has a point and needs an eye.*

**Tom Gilbert**

I'll just give you one example of lousy information hidden in what is considered to be great data. The president of a large machinery company was weekly given a huge stack of data in which he could easily spot the taxes on his service truck tires—$540. What did he need to know this for? But search as he might, he could not distinguish the profitability of his used machinery. Everyone thought that used machinery was highly profitable, but once we reduced this nonsense to two pages of what he really needed to know, he discovered that used equipment was a big loser and he could immediately see why.

This is an exception? Certainly not. His company was cited by Tom Peters in his book *In Search of Excellence*, as one of the great companies and Harvard Business School had called its *information* system a model to be copied by others. Wow! Such nonsense.

Incentives? Even worse, if that is possible. People are paid to come to work, not to produce. Top clerks who produce half again what the average produce are paid about three percent better on an average. What does that mean? The top clerks are subsidizing both the average and the low performers. You might think that they must be stupid. Managers talk about the importance of motivation, but they are

# Human Incompetence

talking about the motives people have in them, not the incentives they provide. And when I say *incentives,* I'm talking about money, not dumb things like special parking facilities or posted recognitions. There are those who say that incentives aren't important, but they are talking about those dumb things. Ask top management to settle for an occasional free vacation for themselves and they will laugh at you. I'll never forget a sign posted by workers which said, "Save the praise and give us a raise."

And instruction. Sure, people sit long hours in training. I remember having to sit in a client's class designed to train telephone repair clerks. The charming instructor went on all morning about the psychology of customers and the importance of answering the phone properly. She was so interesting that she could have made the distinction between bituminous and anthracite coal exciting. Then we came to practice time. We each had a phone on our desks and she rang mine first. But I hesitated to answer it, trying to remember all she had told us. But the only thing she had needed to convey was to answer the phone on the first ring and take charge of the conversation by beginning to ask questions. My goodness—a whole morning. When I talked to her later, she said the curriculum was laid out like that. We should treat training like salt. If a little is so good, that doesn't mean that more is better.

Remember my 80-90-100 criteria? Cut the training by 80 percent or more, bring the average performers up to 90 percent of the best or higher; and realize 100 percent return on your training investment or greater? Oh the billions we could save each year if we would only do this. And the superior results we would achieve! With the ease it grasps good sense, the human head absorbs a mess!

# CHAPTER SEVEN

## SUMMING UP: ON LANGUAGE

There are two things about the English language as it is spoken in America: one good, one bad.

The good one is that we so freely borrow words that are useful to us. This is not so in many other countries—certainly not in France where there are great attempts to keep the language pure, whatever that means. I recall a story about a Frenchman boasting about the purity of his language to an American. "We have so many words that express a thing just right; something you (*vulgar*, he meant to say) Americans don't." Asked for an example, he provided *savoir faire* and said that the Americans just didn't have a word like it.

The American clapped his hands in joy and said, "We do now!"

*Savoir faire* reminds me of a joke I must get off my chest before I continue these musings on language. It seems that a Frenchman, an Englishman, and an American were discussing just what was meant by *savoir faire*.

The American offered his understanding: "If you come home and find your wife in bed with another man and silently walk out, that is *savoir faire*.

"No!" said the Englishman. "If you say *continue* as you walk out, that is *savoir faire*."

The Frenchman shook his head, "You just don't get it. If, as you walk out, you say *continue*, and they can continue, that's *savoir faire*."

But France isn't the only country that is protective of borrowings into its language. We spent a spell in Norway lecturing about human performance, and one of our hosts bemoaned the fact that they didn't have such a nice word as *performance*. "Why not just adopt it?" we naturally asked. "Oh, it's considered so vulgar to borrow from the American language."

Then why don't you do what we did, borrow it from the French?" And then I gave a stab at the French pronunciation: *Pair-for-mahnce*. He thought a minute and his face lit up. "Maybe that will do it!"

We later learned that the Norwegians do have the word *performance* in their vocabulary; it just isn't official; it's considered slang. It means how well you do in bed. My

 Human Incompetence

God! It's hard enough to talk about performance in the workplace without being able to call it by its name.

The bad thing about the American language is our growing disdain for grammar. And I don't mean among the lower classes. The growth of the use of the reflexive pronouns, such as *myself*, is really disturbing. Recently we heard the ABC Morning Show host say something like this: "He did these things better than myself and her."

Twenty years ago I wouldn't have bothered to tell a reader what is wrong with this. But it has become so common among the educated classes that I suppose I must. The correct thing to say would be, "He did these things better than she and I." And don't think that the ABC man wasn't well educated; he is a Princeton graduate.

**Transactional talk**

One of my contributions to the language—and I'm not at all sure that I am the only one who has thought about this—is the identification of something I call transactional words. This helps us use our language a bit more precisely. The word *bean* is not a transactional word—it just sits there. But we terribly confuse the words *data* and *information*. Data just sits there until we use it; but information is transactional. Data gets transformed into information, people can act upon it. We are prone to say, "I gave him all the information he needed, but he didn't understand it." It would be better to say, "I gave him all the data he needed, but it didn't inform him." The first statement puts the blame on

## Confessions of a Psychologist

the employee, the second lays it on management.

We have before us a one-and-a-half-inch document that comes from a manager's department that refers to itself as Information Services. Is this document *information*? Certainly not; it is a huge stack of data that informs him hardly at all. Right at the top of page 127 it tells him that he spends $547 a year on the excise taxes for his service truck tires. Does he need to know that? Certainly not. But nowhere in this mass of data can he find out that his used equipment sales are unprofitable or why, and he surely needs to know that. His warranties on the used equipment are too good and he is paying too much for the equipment on *trade-ins*. That's information he could do something about. But he certainly has plenty of data, piled high and ever growing. What does he need it for? This comes under the name of superstitious behavior. Don't walk under ladders and do have piles of data on your desk and everything will be alright. Not likely!

You've heard this referred to as the Information Age and we hear talk of the Information Superhighway. This is no information superhighway; it's a data dirt road that wanders off aimlessly into the woods and circles around to come back into itself. And, strange for a road, it is wider than it is long. You can't get on it and know where you are going. We are a long way from the Information Age. These are the Data Dark Ages.

*Motivation* is another one of those misused transactional words. When we say our employees aren't motivated, what do we mean? We mean that they lack some drive inside of them. But think about it. The unmotivated pigeon won't

 Human Incompetence

peck the window in the Skinner box. Why? It can be because of two different reasons. First, the pigeon may lack the drive—yes, it is no longer hungry. Or it may be the fault of the incentives we supply it; say the food isn't good. Of course, if you posit the motivational failures within the worker, it becomes their fault. But if you point to the incentives, it becomes the responsibility of management. Where do we usually point the finger of blame? You got it. Right at the employee.

*Performance* of course, is the ultimate transactional word. It embraces both the behavior of the performers and the outcomes they accomplish.

## Vantage Points

*The eye can travel to a star*
*So fast that Science cannot measure;*
*How vast a scope the eye can grasp—*
*Infinitudes of pounding pleasure.*

*And Science can little comprehend*
*What tiny things the eye can see:*
*Imaginary rebuffs of love—*
*Infinitesimal misery.*

**Tom Gilbert**

End of Autobiography

Addendum

# Outline of a Proposal For a Teaching Laboratory

Thomas F. Gilbert

NOTE: This was the formal proposal outline that Dr. Gilbert sent to the Federal Department of Health and Human Services in the mid-1950s for funding. This defined much of his work in Tuscaloosa, Alabama. The grant was not funded as described in the autobiography but the outline of this proposed work is still relevant and of value today.

A. Some Assumptions involved in the development of the laboratory:

> I. Behavior research is concerned primarily with the determination of the optimal conditions under which subject-matter proficiency is attained. The arrangements for schoolroom administration of these conditions are the primary responsibility of schoolroom administrators. For this reason, the teaching laboratory need not be a replica of the schoolroom.
>
> II. The value of any programmed teaching procedure will be determined by the success of the student being instructed by the program.
>
> III. Generally, there are two problems involved in developing subject-matter proficiency in the human: (1) to place the desired behavior under appropriate stimulus control, and (2) to maintain the behavior

through the course of instruction. Basically, both of these depend upon the reinforcement value of the program.
- a. Self-instruction programs will depend very heavily on program-intrinsic reinforcement if aversive control and individual teacher attention is to be minimized.
- b. One of the striking features of existing self-instruction programs is the reinforcement value of frequent success in unfolding knowledge.
- c. Knowledge may be one of the most "primary" reinforcers existent for humans. Young children frequently evidence more knowledge seeking behavior than older children. Some humans will go without food to learn (e.g., university scientists).

IV. The programming of verbal or symbolic knowledge is the programming of behavior. Such behavior follows the same general laws as other operants.
- a. A problem should serve as a conditioned reinforcer if it reliably sets the occasion for successful answering behavior.
    1. To condition people to seek out problems is to produce the appropriate scheduling of successful answering behavior.
    2. If a problem contains good stimulus control over successful answering, it should serve as a reinforcer, and both the question of good stimulus control and behavior maintenance partly rests within the reinforcing nature of problems.
    3. The reinforcing nature of access to problems

Human Incompetence

  is an important part of a teaching laboratory.
V. The success of operant conditioning methods in behavioral analysis has been due to
   a. the use of careful environmental control.
   b. automation of experimental programming.
   c. careful measurement techniques.
   d. intensive studies of individual organisms.
   e. emphasis on control rather than prediction.
   f. emphasizing the question "What variables control?" rather than "How will this variable control?"
   g. use of organisms over whom control can be exerted.
   h. a systematic simplicity rooted in behavioral terms.

VI. In summary, the proposed laboratory would meet the criteria of an operant conditioning laboratory; stay out of the school setting; use subjects whose behavior can be controlled by the contingencies produced in the experimental space and not by instructions or social pressure; supply a means for presenting a variety of teaching programs; and have a facility for investigating the reinforcement value of problems.

B. The Method:

I. Generally, the experimental space consists of an integration of an automatic teaching machine (such as those used by B. F. Skinner) and a human operant-conditioning apparatus (such as used by O. R. Lindsley). Modifications are made where relevant.

a. The human operant-conditioning laboratory consists of a panel mounted in an otherwise barren room. On the panel is at least one manipulandum, a reinforcement tray, and at least one stimulus window when needed. Reinforcement in the form of money, cigarettes, etc. is made contingent upon the operation of the manipulandum. This apparatus allows for the investigation of human behavior from simple conditioning to complex discrimination and chaining.
b The teaching unit is a modification of this experimental space. The attached figure represents a rough design of the mock-up now being built.
   1. A problem screen for projection of problems by photo-slides
   2. A shutter to cover the correct answer
   3. A money, cigarette, etc. reinforcement tray
   4. One or more Lindsley manipulanda
   5. *Right and wrong* answer buttons
   6. A speaker for presentation of auditory stimuli
   7. An answer reception window, written answer reception device interchangeable with multiple-choice buttons or word-number construction sliders
   8. An additional stimulus window for presentation of such things as cumulative counts of student success, prompt variations, etc.
c. The programming of the apparatus is done in the manner typical of operant conditioning laboratories.
d. The peculiar feature of the apparatus is that

 Human Incompetence

presentation of teaching machine programs can be made contingent upon the operation of the manipulandum.
   1. Separable contingencies can be provided for problems, answers, and additional information.
   2. Program variations can be made contingent upon the selection and operation of different manipulanda.
   3. Programs and other reinforcers can be made separately contingent upon different manipulanda.

C. Examples of a few classes of problems that may be investigated by the method.

I. As a teaching machine:
a. The experimental apparatus consists of an automatic training device with many possible modifications. Problems investigated by any teaching machine can be investigated here.

II. The reinforcement value of a program.
  a. Conditioning and extinction: The reinforcement value of a program: The presentation of one or more problems made contingent upon operation of the manipulandum.
      1. The comparative reinforcement value of programs written according to different programming principles, e.g., difficulty of program steps
      2. Separation of reinforcement value of

problem and answers
3. Growth of reinforcement value in time
4. Variations in length of magazine cycle (number of problems available per reinforcing occasion)
5. Reinforcement value of non-programmed materials such as text, pictures, etc.
6. Schedule effects of program items, answers, etc. (including pacing)
7. Investigation of differential conditioned reinforcement value of parts of the problem answer chain:

B4 (manipulandum) − srD (problem) − B3 (answering) − srD (access to correct answer) − B2 (obtaining correct answer) − srD (correct answer) − B1 (obtaining new problem).

8. Discovery of means to increase generalized reinforcement value of prolems ("conditioning" interest in knowledge)

b. Comparative reinforcement values; allowing two or more events to be contingent separately on two or more manipulanda. Student free to select event.
1. Comparative value of a program written in different ways
2. Comparative values of different subject-matters
3. Change in comparative values with time (e.g., how to increase interest in mathematics)

4. Comparison of value of programmed and textual materials
5. Comparisons of programs with other reinforcement, such as money
6. Comparative values of programs where answers are shown on diffeent schedules (e.g., contingent event may be problem; problem and answer shown together; problem, answer shown after answer is made; or problem, answer shown on every $n^{th}$ occasion of answer made)
7. Possible even to calibrate reinforcement value of program with other reinforcers

c. Self-evaluation of students:
1. Working for additional prompts (using additional stimulus window). Student pulls plunger to obtain additional prompts. Degree of prompting necessary to be determined by student.
2. Student may be allowed to skip parts of programs as "too easy."
3. Student may be allowed to "cheat" experimentally.
4. Differential contingencies on above behavior: (e.g., too many "too easy" judgments followed by errors produces time-out or backtracking of program; too few "too easy" judgments may produce similar contingencies. Device may be a "confidence trainer."
5. Self-testing or review may be programmed to student work.

d. Satiation, deprivation, and facilitation
   1. Satiation effects of subject-matters, of item class repetition, etc.
   2. High reinforcement value program may be substituted by other programs
   3. Inter – and intra–variability required to produce minimum satiation
   4. Facilitation and suppression effects of extraneous stimuli, instructions, etc.

III. Other functions of the laboratory.
   a. Such a laboratory could serve as a "program-screener" to screen teaching programs in a way roughly analogous to the "drug screening" procedures.
   b. Investigations of education as a therapeutic agent
      1. With juvenile delinquents who are said to be that way because of repeated failure particularly academic failure
      2. Acute schizophrenia, hospitalized alcoholics, psychotic children, etc.
      3. Problem children in schools
      4. Shaping more problems seeking and solving behavior wherever social illness is characterized by a minimum of this behavior.
   c. Special educational investigations
      1. Problems of instructing the gifted child
      2. Problems of instructing the blind
      3. Problems of instructing the feebleminded.
      4. Problems in revealing subject-matter aptitudes hidden by typical educational procedures
      5. Problems of preparation of institutionalized delinquents, prisoners, etc.

# MEMORIES OF THOMAS GILBERT

Following are the remembrances of Thomas Gilbert by a few of the people who were influenced by and/or who were friends of Tom Gilbert. They are listed in alphabetical order:

| | | |
|---|---|---|
| Bill Abernathy | Joe Harless | Byron Menides |
| Roger M. Addison | Jan Holmstrup | Laura Menides |
| Carl Binder | Kent Johnson | James S. Moore |
| Dale Brethower | Joy Kreves-Yavelow | Tony Moore |
| Phil Chase | Yuka Koremura | Richard M. O'Brien |
| Bob Cicerone | Danny Langdon | Carol M. Panza |
| Ellen Freed | Joe Layng | Caroline Sdano |
| Joan de Haas | Marie E. Malott | Julie Vargas |
| Mark de Haas | Stuart Marguiles | Kathleen Whiteside |
| Dan Hardin | John McKee | |

*Not everything submitted was included but an effort was made to ensure Tom's impact on each contributor remained clear.*

# Bill Abernathy, Ph.D.

*Author*

*Assistant Professor of Psychology,
Southeastern Louisiana University*

*Founder, Abernathy & Associates*

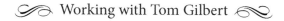 Working with Tom Gilbert

I first met Tom when we jointly consulted at Union National Bank in Little Rock, Arkansas. Tom and I were initially hired to develop a teller cross-sell training program for the bank. In our first meeting with bank management, Tom pointed out that the teller cross-sell was likely only one of many improvement opportunities across the bank and that it might not be the most profitable. Management agreed, and we proceeded to collect data and compute performance improvement potentials (PIPs) across the entire bank. I later employed this method of identifying PIPs with Tom at AT&T and also on my own at European American Bank.

The PIP introduced three important concepts to the practice of Organizational Behavior Management (OBM). First, performance variability is viewed as an opportunity. This view is a significant departure from industrial engineering, which more often finds variability a liability rather than an opportunity. It is usually the case that behavior analysts seek to initially increase performance variability rather than decrease it. Performance improvement is essentially an increase in variability. This distinction between the two disciplines is primarily due to OBM's focus on the individual employee's performance while Industrial Engineering focuses on group averages and variances.

Second, the PIP provides a simple and objective means for selecting target results for improvement. Tom's PIPs were computed on the results of employee behaviors rather than on the behaviors themselves. This, too, is a key difference between Gilbert's approach and some other approaches in OBM. At Union National Bank and AT&T, Tom instructed us to gather existing financial and operational data to compute PIPs. These data were all based on results rather than the behaviors that produced them.

Third, Tom used a variety of improvement strategies once a target result was selected. He termed this approach *behavioral engineering.* He organized potential performance constraints into two sources – environmental and individual. He further delineated each source by the effects of the stimulus, response, and consequences. The stimuli and consequence constraints were similar to the improvement techniques used in most OBM projects. However, the response category constraints were much more akin to industrial/organizational psychology and industrial engineering. Tom believed that many managers were too quick to assume that deficient performance was due to individual constraints. He believed that environmental constraints should be examined before individual ones.

 Human Incompetence

| Source | Stimulus | Response | Consequences |
|---|---|---|---|
| Environment | Information Job descriptions, guides, feedback | Resources tools, time, materials, access, staffing, processes | Incentives: monetary and non-monetary incentives, career development, consequences for poor performance |
| Individual | Knowledge training | Capacity Selection procedures, flexible scheduling, job aids | Motives: Worker's willingness to work for available incentives, recruitment process |

Tom's consulting firm was titled Praxis. The *Random House Dictionary* defines *praxis* as practice, as distinguished from theory, application or use, as of knowledge or skills. This suited Tom who in my experience was much more interested in a behavioral technology than in behavioral theory. Tom was a problem solver. In our nightly discussions after work, Tom would excitedly describe novel solutions he had developed for training issues. He described dozens of successful applications over the course of many evenings.

For example, beginning clarinet players typically fail to place their fingers correctly on the holes in the clarinet. The results are squeaks or wrong notes. Tom noticed that many beginning players lifted their fingers too high and thus missed the proper placement when they lowered their fingers. Tom's simple, but unique, solution was to string a wire down the top of the clarinet to prevent the players from lifting their fingers too high. Tom developed a number of such techniques for simplifying the learning of musical instruments. As a musician myself, I was very impressed with his solutions to enduring problems for beginning musicians.

Tom was a student of Greek mythology and literature.

He would wax eloquently for hours about these and other diverse topics. Often, he would synthesize disparate ideas from many sources to make a point about behavioral technology and theory. We all miss Tom and his keen insights and humor. I was honored to have the chance to work and converse with him for several months.

# Dr. Roger M. Addison, CPT

*Senior Director of Human Performance Technology, International Society for Performance Improvement*

## ～ Tom Gilbert ～

I first met Tom Gilbert when I was nineteen. I was working at TMI (Teaching Machines Inc.) Learning Lab in Albuquerque, New Mexico. Tom came to visit Lloyd Homme and Jim Evans. Lloyd, Jim, and Tom had met in graduate school and remained friends for many years. Over the years our paths crossed many times. Tom had a bigger than life reputation, and many stories have been told about him. Some are true and many became better with age.

Tom's book *Human Competence* had a great influence on my thinking about improving individual performance. As we developed the current ISPI Institutes we used his six-cell model as part of the history of Human Performance Technology.

We have asked over two thousand Institute participants to determine how they could improve their own performance, and as you can guess, it is not the individual performer, but rather the environment, that is most often in need of improvement.

I last saw Tom at an ISPI Conference, and as always he was very charming. By then he was one of the senior statesmen of ISPI and many came by to pay their respects. However, my best memories of Tom, along with Lloyd and Jim, are of both professional and personal adventures.

# Carl Binder, Ph.D.

## *Co-founder, Binder Riha Associates*

 Tom Gilbert – My Unforgettable Character, Indeed!

I was privileged to know Tom Gilbert from the mid-1980s when I was Program Chair and then President of the Boston Chapter of NSPI – the National Society for Performance Improvement (which became ISPI a few years later) – until his passing in 1994. I was responsible for growing the Chapter by attracting new members using innovative programs and workshops, an effort that more than tripled membership over a two-year period. I had read Tom's book, *Human Competence*, shortly after it was published in 1978, and was attracted to his work because it showed me a compelling path from the work of B.F. Skinner, with whom I had the pleasure of studying at Harvard during my doctoral work in the 1970s, to "the world of work," as Tom so aptly called it.

The key for me was his identification of accomplishments (what I now call *work outputs*), the valuable products of behavior that contribute to organizational results, as more important than the behavior that produces them. Instead of focusing on the key-peck (behavior) in the prototypical operant conditioning pigeon experiment, Tom recognized that it was the switch-closure (accomplishment) produced by that behavior that actually counts. Or as we say about humans at work, it is valuable accomplishments that organizations need from their people and that link behavior to business results. He also recognized that, in contrast to pigeons who peck keys that are merely effective transducers of behavior, humans are greatly influenced by what he called "instruments" (environmental design, tools, etc.) that support or obstruct behavior, depending on their design and configuration; and that unlike white pigeons or hooded rats, individual humans bring individual profiles of "capacity" to their jobs, and therefore must be selected and assigned based on those profiles. With a focus on accomplishments plus the simplicity of his Behavior Engineering Model, Tom had provided a simple but powerful paradigm shift that made far more sense to me than merely adapting the use of antecedents and consequences to the management of behavior in organizations.

When I invited Tom to deliver a workshop for our local chapter, he responded enthusiastically and arrived in Boston with Marilyn, his wife and writing/thinking partner. The workshop was at times a bit disorganized, due to Tom's tangential harangues about the "cult of behavior" and other such ideas. But the substance was brilliant and I came to know him and Marilyn as friends and mentors, a relationship that continues to this day with Marilyn, who lives a few miles from my home on Bainbridge Island, near Seattle, and who continues to amaze me with her own vitality and brilliance.

# Human Incompetence

Over the years I saw Tom at conferences, always learned something from him, observed his extreme brilliance, experienced his outlandish humor and challenging provocations, and also noticed the Southern Gentleman just beneath the surface of the iconoclastic rabble-rouser who was always a little frightening because of his unpredictability. I heard directly from him such stories as when he drove his rented car into a river, but escaped the sinking vehicle in the nick of time by slipping through the open window, despite the electric window motor's failure. I heard from Og Lindsley, Tom's old friend from their shared years at Harvard, about late-night carousing and competitive brilliance; and from Tom's longtime protege and colleague, Joe Harless, who was subsequently a mentor to me, about even more of his brilliant and outrageous behavior. So I was well-prepared for just about anything that might come from Tom Gilbert, brilliant or otherwise.

An event that stands out in my memory occurred toward the end of Tom's life, and illustrates both his brilliance and his sometimes outrageous conduct. As an aside, I have always been amazed by the stories I hear of pioneering performance technologists and behavior scientists in the generation just before my own. Not just Tom, but so many others seemed able to drink more alcohol, smoke more cigarettes, and party harder than anyone I could think of in my own generation, excepting perhaps a few rock bands. It sometimes seemed that these men, and the occasional women who could drink and stand on their own with the guys in the "man's world" of their day, did their most creative work while three sheets to the wind. Somehow, the toxins didn't seem to dull their brilliance. As with a few

## Confessions of a Psychologist

legendary poets and novelists, it might even have made a positive contribution!

My story is from an ISPI conference in the early 1990s. I was then Chairman of Product Knowledge Systems, Inc., a company that I co-founded with Information Mapping, Inc., and which served Fortune 500 sales and marketing organizations by helping accelerate the results of new product launches and marketing strategies using performance improvement methods. At the time, our largest client was a well-known biotechnology firm, a source of million-dollar contracts that kept two dozen of our consultants and specialty staff occupied. I and one of my account managers were hosting a pair of senior people from the client's Sales Development Department, directing them to relevant conference sessions, sharing meals and hospitality suites, and introducing them to our friends and colleagues in the field.

One evening before dinner we ran into Tom Gilbert and an old colleague of his, Jim Evans, in the lobby. Jim was long-retired, attending the conference in overalls and a beard that made him appear more like a farmer than a brilliant instructional technologist. Despite appearances, Jim and Tom engaged in repartee that was partly crazed and partly brilliant, a combination I had come to expect in Tom's company. To impress my client and enjoy time with esteemed gurus, we invited them to dinner, and proceeded to have a most interesting conversation, alternating brilliant suggestions from Tom about improving sales performance and revolutionizing training with orders for more drinks. After a while, I could see that my client was getting a little concerned at how much our guests had imbibed, and how wild their con-

113

 Human Incompetence

versation had become. Nonetheless, we marveled at Tom's continued brilliance. At some point I moved to graciously end the dinner, claiming the need to get some work done in my hotel room and hoping to end it before Tom and Jim became too outrageous or loud. We excused ourselves, bid one another good night, and retired for the evening.

The next morning I saw Tom in the lobby, looking chipper and none the worse for the evening's celebration. I noticed that Jim was not at his side, and asked how the rest of their evening had gone. Tom explained that things had been going very well when Jim climbed up on a marble pedestal in the lobby as they continued their verbal play. At that point, Tom told me, Jim had fallen off the pedestal, "cracked his head" and had to be taken to the hospital. A mixture of dread and guilt came over me, with the feeling that I had been responsible for a terrible accident by buying them drinks the night before, and not stopping earlier. I said to Tom, "Oh, I'm so sorry to hear that!" Tom simply responded, relaxed and matter-of-factly, "Oh, don't worry. He's OK. It happens all the time." I was stunned, relieved, and amused all at the same time! Something about the whole scene sums up my recollections of Tom, the brilliant iconoclastic madman.

Tom Gilbert's contributions to my professional life have been profound. I have always said that his book contains more ideas than most people could ever use in a lifetime, let alone come up with or articulate. For me, the essentials are a focus on accomplishments (work outputs), and viewing the factors that influence behavior as a combination that must be coordinated, tuned, balanced, and understood systemically. I give Tom credit almost daily, since the core of my present work is a descendent of his Behavior Engineering Model plus a visual representation of the Performance Chain, linking business results backwards to work outputs, behavior, and behavior influences. When I first used Tom's Behavior Engineering Model, I found it brilliant and very helpful in what we were doing at the time (the late 1980s).

Over the last twenty years, we have set a goal of introducing research-based methods to corporations, families, and organizations of all kinds – to reach a non-specialist audience with the ideas that originated with Skinner and Gilbert. We always trace their conceptual underpinnings to Dr. Tom Gilbert, one of my most important mentors, and certainly quite an unforgettable character.

# Dale Brethower, Ph.D.

*President, International Society for
Performance Improvement
Professor of Psychology, Western Michigan University*

## Tom Gilbert, the Man

Tom Gilbert was articulate, red-haired, and hyperactive the first time I talked to him in 1959 and the last time I talked with him a few months before his death thirty-five years later. We originally met when he was a newly minted Ph.D. and I was a graduate student. We attended meetings, hosted by B. F. Skinner, which were attended by researchers from the greater Boston area. Little did we know, in 1959, that two things would happen over the next few decades: 1) The International Society for Performance Improvement would recognize Tom's contributions by naming one of their highest honors after him; 2) I would be honored to receive The Thomas F. Gilbert Distinguished Professional Achievement Award.

Tom and I got much better acquainted several years later at the University of Michigan when he came to Ann Arbor many times to work with Geary Rummler and others, including me, at the Center for Programmed Learning for Business. Tom was always creative, a good friend, and a harsh critic. He didn't abide fools or pretentious people but he was kind and considerate of his friends. For example, Tom, Geary Rummler, and I were having dinner in the Rubaiyat restaurant in Ann Arbor. A distinguished visiting colleague began expounding on the importance of *float* in financial management. Tom noticed that I looked blank and saved me from embarrassment by a five-minute harangue at the poor man about people who use jargon without defining their terms.

I tell these stories to illustrate that, as you the reader may surmise in these stories of Tom's life and his impact on us all, Tom Gilbert, while brighter and more creative than most people, Tom was a fallible human being, both vexing and loved.

# Phil Chase, Ph.D.

*Director, Cambridge Center for Behavioral Studies*

## Tom Gilbert
### ○◦ Two memories, one indirect and one direct ◦○

Tom Gilbert was one of the three most influential intellectual leaders of my life and I have two memories to share concerning his influence. The first is indirect, an effect of his writing. I never taught an applied course that did not include something of Tom Gilbert's in it. I often quoted him; I copied his ideas in slides; I assigned his works as reading; I set up discussions to argue his most outrageous suggestions. So not only did he influence me, he influenced my students. The second is direct. Years ago Aubrey Daniels hosted a bunch of us at his office during an ABA meeting in Atlanta. We went down into a training room to discuss OBM and especially how academic programs and behavioral consultants could best work with each other. Tom was there. When the topic came up concerning how consultants could offer internships to graduate students, Tom stood up leaning on his cane and said, "I would love to have interns, but I don't know what they would do. I suppose they could watch me lying on a couch because that is where I do most of my work." It occurred to me, even though I was no longer formally a student, I'd have given my eye teeth to be able to interact with Tom every time he got off that couch.

# Bob Cicerone, Ph.D.

***President and Consultant,
Performance Systems for Success, Director
of Customer Loyalty, ETC Institute***

## 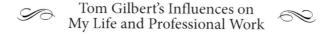 Tom Gilbert's Influences on My Life and Professional Work

Tom Gilbert profoundly influenced my personal life and my professional work. Some key examples are described in the following paragraphs.

Tom's decision to hire me from a university faculty position created my first step toward a career in industry. Before I joined Marilyn and Tom in 1980 as an associate consultant in their firm, Performance Engineering Group, I was a tenured member of the faculty of Montclair State University (NJ). In 1978 I decided to leave academe for industry. As I searched for employment in industry during the next two years, I was frequently told that because I was a teacher and held a doctoral degree I could not do anything practical and useful for a business and, therefore, I should resign myself to remaining a teacher. Fortunately, my teaching experience and doctorate did not deter Tom from hiring me, perhaps because our histories were similar.

Tom's decision not to use the Behavioral Engineering Model with his associates in Performance Engineering Group created many non-preferred experiences for me. He did, however, give me advance warning. When I joined Performance Engineering Group, Tom told me that he did not use within his company the performance engineering tools that he created for client organizations. Unfortunately, he was true to his word.

---

The best years of my professional career (forty years and still counting) are due to Tom. He was forced to terminate my employment when the contract that had enabled him to hire me was unexpectedly and drastically reduced in scope by the client company. A few months later, Jack Zigon, who had recently been selected by Yellow Freight System to start a training department, asked Tom to recommend potential candidates for an internal consulting position in Jack's department. Tom suggested that Jack contact me. Tom's referral ultimately resulted in Jack inviting me to join his department. The experiences I had while being managed by Jack and, later, when I managed the department were, by far, the most highly preferred experiences of my career because of the value of the accomplishments we produced by using instructional systems design and performance engineering systems, both within the department and to deliver services to the company. Zigon and Cicerone (1986) describe an example of our work that explicitly shows our use of tools that Tom invented.

I first learned about the value and importance of instructional objectives from Tom, as well as how to create instructional methods that effectively achieve learning objectives. Tom was also the first to impress upon me that managers are responsible for proactively engineering their subordinates' successful job performance. As a manager I've used tools that Tom invented, such as the Behavioral Engineering Model and job models to engineer the job performance of my subordinates. In addition, the process I developed for managing customer satisfaction and loyalty (Cicerone, 2009; Cicerone, Sassaman and Swinney, 2005; Cicerone, 1998) includes the Behavioral Engineering Model as a key component. I've also used the Behavioral Engineering Model to create a non-verbal process for managing the behavior of motor vehicle drivers (Cicerone, Sassaman and Swinney, 2006) and used this process during field research in which drivers are intercepted and interviewed while they are en route to their destinations.

---

Tom's ideas about the critical role that theory plays in guiding a search for the reasons an opportunity exists to improve a valued outcome caused me to reverse my evaluation of theory. When I completed the doctoral degree in 1971, I agreed with Skinner's argument (e.g., Skinner, 1950; Skinner 1963) that too little was known about the actual biological and environmental causes of behavior to justify the all-encompassing theories that then dominated Psychology. Tom's ideas on the value of a theory to generate trouble-shooting questions that can be used to find the reasons for unacceptable outcomes caused me to change my evaluation of the role of theory (see Cicerone, 2009; Cicerone, Sassaman and Swinney, 2005).

# Ellen Freed

## *Training consultant and marketing writer for the arts*

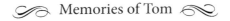 Memories of Tom

I first met Tom Gilbert in the spring of 1968 when he hired me as a writer and junior instructional designer at Praxeonomy Institute—the precursor of Praxis Corporation. At this time, Robby was almost two, and Eve was three months; and Marilyn presided over an extended family that included a total of eleven children. Since Tom and Marilyn lived around the corner from the office, I was often invited for dinner.

After being directly trained by the master in mathetics, my first assignment was to develop a course plan for teaching children how to play the clarinet. Other educational projects included beginning reading and social studies. It was both a privilege and an opportunity to work directly with Tom on these programs—to experience firsthand the power of his intelligence and creativity. Indeed, with every subject, he injected instructional design ideas that made learning more effective, engaging, and fun. To this day, education is in desperate need of the innovative thinking style of Tom Gilbert.

Thus began a close professional and personal relationship

 Human Incompetence

with the Gilbert family that has lasted to this day. Selected memories of Tom follow …

## April 1969: Conference Room, Praxeonomy Institute*

In a professorial pose, cigarette in hand, shirt sleeves rolled up, Tom stood at the blackboard ready to work! We were there, master and adoring disciple, to develop the mediation system for the "Mathetical Beginning Reading Series." This was a program commissioned by Lee Brown to teach preschool children to read and write in three months.

Tom's creativity was flowing as he presented his first competitive grouping of mediator candidates—five or six beauties in a row—the first members of a picture alphabet designed to facilitate both encoding and decoding. Each mediator had to be easily identified by four or five-year-olds, (such as "apple"), have the basic shape of the letter (a) and begin with the sound associated with it (aaa). There was an ever-increasing energy that emerged as he drew each image on the blackboard, and the originality of his thinking seemed to light up the room.

Stopping for a well-deserved rest, Tom said: "Ellen, do you have any ideas?"

I had not expected to participate as a peer in this high-level brainstorming session, and Tom's confidence in my ability to contribute almost jolted me out of my seat! I remember sitting up straight, and, as if in a trance, I produced two or three acceptable mediators. It was as if Tom's inventiveness had somehow been transmitted directly to me, and I re-

Confessions of a Psychologist

*Robby's Wedding, The Gilbert's Home, Hampton, New Jersey*

ceived it just in time.

"Good," he said, after which he proceeded to complete the entire mediation system on his own in about an hour.

With the generosity only true geniuses can afford, Tom Gilbert gave me the gift of my own creativity and competence on an unforgettable afternoon, sometime in 1969.

*I wrote about this memory in the Tribute Edition of *Human Competence*, by Thomas F. Gilbert, published in 2007 by Pfeiffer, Imprint of John Wiley & Sons.

## November 24, 1989: The Gilbert Kitchen in Hampton, New Jersey

Thanksgiving at the Gilbert's charming country home set on five acres of land had become a tradition. One could count on great food, a beautiful table, and most importantly, good company.

Tom, a gourmet cook, was in the kitchen supervising the preparation of the meal. I remember standing next to Eve who was stirring a pot of soup. She was nine months pregnant and about to give birth to Tom's first grandchild any day. He was pacing back and forth, checking the turkey,

 Human Incompetence

tasting the stuffing, adjusting the seasonings where needed. This was Tom—the family man who would soon be surrounded by most of his eleven children, selected friends and colleagues at a warm table—Southern hospitality in the Northeast.

P.S. Georgia Gilbert was born three days later. In April 1990, I attended a presentation given by Tom at an NSPI conferenceHe began by telling the audience about the arrival of his first grandchild. As he spoke, I remember thinking that I had never before seen Tom radiate such joy.

## June 21, 1995: Robby's Wedding, Hampton, New Jersey

It was a beautiful summer day, and the ceremony was held outdoors. Tom was overcome with emotion, trying unsuccessfully to hold back tears as Robby and Juliette took their vows. I believe this side of him was unknown to most of the people with whom he interacted. However, for me, it represented a bond between us that was consistently felt, but never discussed. I could look at him and know we shared membership in a psychodynamic travel club, having both journeyed to some heavy emotional territory.

This unspoken communication was always there whenever I saw Tom in the years following my departure from Praxis Corporation in 1971. And, on this joyous day, after the ceremony when the guests were enjoying good food and drink, I looked for Tom. He was nowhere to be found. I walked around the grounds for several minutes, and finally

I saw him sitting alone on a bench at the back of the house. His eyes were red and his face bore the evidence of the deep feelings he had just experienced. I took his hand and sat with him in silence ...

## September 1995: Nursing Home, Clinton, New Jersey

Irving Goldberg, Stu Margulies, and I went to visit Tom to say goodbye. There he was—this amazing man—lying on a bed with tubes coming in and out of him. He was dying of cancer, but even in a prone position, Tom stood tall.

I kissed his forehead, and he uttered these last words to me: "Ellen, you are a very practical person, and don't you ever forget it!" Coming from Tom for whom practicality was a high value, this was tantamount to the Good Housekeeping Seal of Approval. As he was approaching death, he found the energy to give me a gift I shall always cherish.

## October 11, 2009: Upper East Side, New York City

I can close my eyes and see Tom standing with that erect posture of his—a complicated man with many talents and the creativity and intelligence of a true genius. He loved his work, his family, country music, good food and drink. Amazingly, he functioned competently, creatively, and innovatively while containing a huge reservoir of emotional pain. He leaves behind a wife of thirty-five years whom he adored, numerous children and grandchildren, a body of work, still revered and used, and many friends and colleagues who remember him with love and admiration.

# Joan de Haas

*Neighbor, friend, artist*

# Tom Gilbert

The first time I saw Tom Gilbert was on a rural lane in Hampton, New Jersey. As I walked to the mailbox, a car with Tom and Marilyn approached me. The car stopped, and the Gilberts introduced themselves and told me they had bought the old white farmhouse just farther down "our" lane.

They both were so gracious and gregarious that I "took" to them immediately. Tom always had a smile on his face. And whenever he met me, he would tell me I was the second most beautiful woman he ever met because Marilyn was the first most beautiful woman he ever met. He was quite a charmer and always had the funniest sense of humor. He was the intellectual's version of W.C. Fields! Later on, when he needed to use a cane, he even wielded it around like Fields did.

One day, Tom, Marilyn, my husband Siegmund, and I were at the Barnes Foundation in Philadelphia. We enjoyed the great masterpieces hanging on the walls and of course commenting on the great art collection. If you know the Barnes, it displays its treasures the way they were displayed in the late nineteenth century. Everything seems to be on top of each other, and there is nowhere one can move without looking at where you are going. Well, there was Tom, commenting and waving his cane around in the tight quarters, as though it were a pointer in the classroom. I thought the guard was going to faint before I stepped forward between the cane and a large Matisse painting that seemed in peril! I mentioned to Tom it probably would be better not to bring the cane up any higher than ten inches off the ground while we are at the Barnes, because if you break it you bought it, sort of.

There are so many images and memories of Tom that bring a smile to my face, because he was a genuine human being with a great deal of love in his heart. He so loved Marilyn, and his children and grandchildren.

Tom thought that science should make the complicated simple. Tom had depth, humor, intelligence, and a "good core"; it was a joy to know him and his wonderful family.

# Mark de Haas

## *CFO, Fairstar Heavy Transport NV*

I remember when the Gilberts moved into our neighborhood. I was in college but came home on weekends, and for our small community nestled in the woods of Hunterdon County, a new neighbor was a big deal. My family quickly became friends with the Gilberts because of our similar interests in the arts and literature, and they were just damn fine people. Tom's wit and fun nature could win over anyone, and it was always a pleasure to talk with him and just hang around visiting. Some of the best memories I have of Tom are when we would sit out on the back porch and talk about philosophy, art, music, how to smoke a turkey, and at times—performance engineering. Tom had a very practical outlook on life, and yet he lived in a world of ideas and theories. His mind was always working, and he would comment on something with a quick burst of insight and I can still hear his voice now. We would work late together, and as he would talk and write notes, I would translate his concepts to computer code. Tom gave me my first technical job as a software engineer, and I remember those times fondly. I never felt like an employee of Tom's, but rather a close friend and neighbor.

Confessions of a Psychologist

I remember the time Tom asked me to come down and look at his new lawn mower. It simply would not start, and after a while of pulling on the cord we discovered that Tom had put the oil in the gas tank and the gas in the crankcase. In his typical humor, he said something to the effect of "How could YOU have done something so stupid!" We all laughed. Tom and his family will always have a special place in my heart, and I cherish the memories of all the good times we had together.

# Dan Hardin

*President, Seattle Chapter of the International Society for Performance Improvement*

 A shift from instructional designer to performance technologist

Like most performance technologists I have met, I came to the field of performance technology as an instructional designer.

As a Coast Guard Aviation Survivalman I was assigned in the early 1980s as an instructor and course designer to assist in the development of a training course. As a subject matter expert (SME) I was not familiar with instructional design. I was one of a staff of three assembled to establish this new course. Not knowing how to develop training we were lucky to be introduced to a systematic method for developing instruction. Led by our Education Specialist, Charles Swarengen, we were introduced to the instructional systems design (ISD) process. We completed a front end analysis using what we learned from Joe Harless and developed training objectives as prescribed by Robert Mager. We developed tests that matched the objectives, then developed training to enable the students to accomplish the objectives.

Kirkpatrick's four levels of evaluation were adopted for our evaluations. It was great to have a systematic method for developing instruction that was based on the requirements necessary for performance on the job (the real world). Being systematic, the process of developing our course was relatively easy.

After attending a Harless Job Aids workshop, I became obsessed with developing powerful job aids as an alternative to training for some job tasks. It was obvious job aids were very powerful. Where training is used to help performers store information in their memory, a job aid is a storage place for information other than human memory where the information needed to accomplish job tasks are stored in the environment on the job.

A few years later, I was given a book to read; Dr. Gilbert's book, *Human Competence: Engineering Worthy Performance*. All it took was the first three chapters of this book to put everything into perspective for me. In the first three chapters Gilbert built a foundation for creating what he called the Behavior Engineering Model (BEM). This simple six-cell model clearly puts strategies for affecting human behavior in a common sense matrix. Each cell of Gilbert's Behavioral Engineering Model presents varying degrees of leverage in its potential to improve performance; this explained why job aids were so powerful compared to training. Job aids fit in the first cell which has the greatest leverage. Training sits in the fourth cell and, although powerful, has much less leverage when talking in terms of the potential to improve performance for the least cost. Sure, training is a powerful intervention but it is a costly undertaking compared to placing the information needed to do the job at the job site. It's all about leverage.

From the day I read *Human Competence*, my world as an instructional designer expanded to the world of performance technology. As a result of reading this work, I have continued to hone my skills as a performance technologist. I enrolled and completed a master's degree in Instructional and Performance Technology at Boise State. I lead a team that developed an electronic performance support tool that was awarded the International Society for Performance Improvement (ISPI) Award of Excellence for outstanding performance aid. I am the president of a local Chapter of ISPI; I have attended various IPSI conferences and have presented at various conferences and other HPT venues in an effort to spread the word about the power HPT has in affecting human performance. At one ISPI conference Dr. Gilbert even signed my tattered copy of his book for me. I continue to refer back to the simple message of those first three chapters. I believe every manager of people would be greatly empowered with this knowledge and I try to communicate that to as many managers that will listen. I often wind up lending out my copy of the book and say, "Just start with the first three chapters. But be careful, it could change your life too."

# Joe Harless

*Author, Accomplishment-Based Curriculum Development System*

## Memories of Tom Gilbert

No other single person has had more of influence on my work than Tom Gilbert. His influence began when I was a psychology student of his in the 1960s. His influence continued over the next few years as I worked with him on a series of landmark projects for the Centers for Disease Control and the Rehabilitation Research Foundation. Working with Tom always inspired an awe of his quick mind coming to grips with problems and generating solutions I had not thought of. His designs were invariably elegant and almost always correct.

Although our consulting firms (Harless Performance Guild and Praxis Corporation) were later fierce competitors, Tom never hesitated to share his mighty intelligence with me — sometimes this "sharing" took place via phone calls in wee hours. Clock-awareness was not one of Tom's strengths.

Like the diagnostic genius physician on the TV show *House*, Tom was not driven by ego or money. Tom was driven by a passion for figuring things out. Even today when I'm addressing a knotty problem, I often have an imaginary discussion with my old mentor. His influence continues beyond his demise.

## Jan Holmstrup

*Fundraising and publicity, Anderson House, Inc.*

I knew Tom during the last of his nine lives. I tried my best to comfort Marilyn at his bedside in the early morning hours after he died. But, is there any comfort to be had at a time like this? Perhaps the serenity on his face, an end to his pain and a lifting of the drudgery he endured as his world was reduced to the confines of a hospital bed?

I imagine Tom's earlier lives were lived in a world that had no bounds; a thrilling, stimulating world full of discovery, adventure, vitality, and accomplishment. At times, it was hard to reconcile the kind, modest, quiet man I knew with the man credited with so many weighty achievements.

Yet, whether I was sharing a meal with Tom, Marilyn, and Georgia on their porch overlooking the hills, riding in the backseat on a Sunday drive through the countryside or watching the election returns at one of the Gilbert's legendary parties, Tom's extraordinary talents couldn't be missed. He never stopped observing, asking questions, pondering, and offering intelligent commentary all his own. I admired how at peace Tom seemed even during difficult times, how graciously he accepted his fate, how fearless he appeared.

Tom had a huge capacity to love. He often shared fond memories of his children, with incredible joy and pride, and of his mother, who was a successful entrepreneur way ahead of her time. His daily companions, Marilyn and Georgia, filled his heart, sustained him, and eased the burdens of his final days.

How fortunate I feel for having known Tom, even just for a small slice of his last life. He probably didn't know how much comfort and confidence he gave me, how welcome and at home he made me feel, and wouldn't easily accept credit for enriching my life. Thank you, Tom. And Marilyn, thank you for asking me for a ride that snowy day in 1988. You opened up a whole wonderful world to me.

# Kent Johnson

*Founder and Director, Morningside Academy, Seattle*

I remember first hearing about Tom Gilbert in graduate school in the 1970s. Organizational behavior management was a growing interest group and career track among behavior analysts. The buzzword at the time in OBM circles was the PIP: the performance improvement potential. I recall this being the first time I heard about measuring the discrepancy between current performance level and a standard or criterion to reach. This simple but elegant concept and measure intrigued me enough to buy a copy of Tom's new book, *Human Competence*: (Gilbert, 1978, 2007), recently reissued.

The next time I was introduced to Tom's work was from a graduate school professor, an ironically antibehavioral one, who introduced me to Susan Markle's programed instruction text, *Good Frames and Bad* (Markle, 1969). I diligently studied that program and later, Markle's subsequent programed worktext *Designs for Instructional Designers* (Markle, 1990), which reorganized and broadened her instructional design methods. In those worktexts, and in our subsequent conversations, Markle taught me Tom's technology of teaching that he called *mathetics*. Mathetics is really the generic roots of Engelmann's popular Direct Instruction (DI) model of scripted programs in basic skills for groups of learners (Engelmann & Carnine, 1982). By the time Engelmann published his first DI programs in 1964, Tom's work was readily available. It appears that Engelmann was heavily influenced by Gilbert's model.

Using mathetics, teachers demonstrate or model expected performance, prompt and otherwise assist students to engage in the performance, and, when the time seems right, they release or test the student's performance. These steps are iterative, based upon a learner's performance, and were originally outlined as a guide to plan programed instruction for individual learners.

I began to examine DI from the point of view of mathetics, discovering that it greatly increased my understanding of DI programs and how to improve their effects on learners. Based upon how students were performing in a DI program, does the instructor need to add more "demonstrate" (priming) tasks to the script? Add more "prompted" and otherwise assisted tasks? Add more opportunities for independent practice through "release" tasks? What decision rules could we develop to facilitate those adjustments? Later, Tom's wife and colleague, Marilyn Gilbert, shared the two and only issues of Tom's *Journal of Mathetics* with me, chock full of ideas for mathetical design in a wide variety of disciplines (Gilbert, 1962a, b). By the late 1990s I was regularly including detailed exercises and discussions about mathetics

 Human Incompetence

in my DI workshops for teachers: how to use mathetics to zoom out to a broader look at DI performance, and how to zoom in from mathetics to Engelmann's more specific version, *Direct Instruction* (Johnson & Street, 2004). I recently had an opportunity to describe the relation between Engelmann's Direct Instruction and Tom's generic mathetics underpinnings to a broad audience in an entry on Direct Instruction that we wrote for the *Encyclopedia of Educational Psychology* (Johnson & Street, 2008).

Tom has gone unnoticed for too long for his brilliant work on direct teaching. He was really the first to comprehensively capture the essence of teaching clearly defined objectives in what I call a generalized imitation training (GIT) model for mastery learning (Johnson & Street, 2004, 2008). He was also one of the first to clearly distinguish when to take a GIT approach, and when other important educational outcomes, such as experiential learning, personal growth, and effective interpersonal skills warrant other approaches.

I actually did not meet Tom until he began attending ABA conferences in the late 1990s. Even near the end of his life, Tom was exploring new and different procedures to improve his teachings and was not satisfied to rest on his laurels. Tom would initiate many conversations with me over the next few years to discuss the fluency-based practice models that were being developed and described by Precision Teachers. He acknowledged the importance of fluency procedures, spoke of this gap in his educational model, and shared many ideas and asked many questions about how he could fill his gap. Tom was truly a listener and learner as well as a contributor in these encounters.

I miss my conversations with Tom. My work as a school director and educational consultant has been immeasurably enhanced by Tom's work, and I owe a great deal of debt to Tom for my work today.

# Joy Kreves Yavelow

*Artist*

## ᐧᣞ The Tom Gilbert I Knew ᣝᐧ

I met Marilyn Gilbert before meeting Tom. She came into my little art gallery in Frenchtown, New Jersey, and we immediately became the best of friends. It is most likely that I first met Tom when she brought him into the gallery. Like so many people, I was amazed at this man who was the utter physical likeness of President Eisenhower, to a degree that could have earned him another job, as impersonator.

Tom frequently accompanied Marilyn to the receptions I held at the gallery. He was a genial presence there and at the gatherings and parties Marilyn so frequently put together at their house in New Jersey. The Tom Gilbert I knew was a warm host and enthusiastic storyteller. It was clear that he relished company, and he always had a story to insert into the conversation – often a story we had all heard him tell many times already—but he so clearly enjoyed the telling, that we all readily listened again. By the time I met Tom he had already published his important work, *Human Competence*, and was mostly retired. His health was breaking down, and periodic diabetic injuries soon began taking a toll. Once Marilyn told me, "I wish you could have met Tom earlier," because in his final years he was evidently not quite the powerhouse he'd been before. He was, however, still an impressive mind.

Tom was also a wonderful caregiver for his granddaughter, Georgia. Marilyn didn't worry about his increasing habit of falling asleep during the day while Georgia was a wide-awake, curious, exploring child, though I did! If we had plans to go out and about, she fully trusted Tom to watch Georgia. I admit my worries proved unfounded, since many times upon our return we would be greeted with Georgia's excitement to show us something Tom had taught her that day. In those years, though on the periphery, Tom's presence enriched my life and it was a pleasure to know him.

# Yuka Koremura, Ph.D.

*Post Doctoral Fellow, University of North Texas*

 Memories of Tom the Man, and Tom's Work

I came to the United States a year after Tom passed away, so I don't have personal memories of him. My first contact with his book *Human Competence* was when I took Dr. Cloyd Hyten's Advanced OBM course in 1997. Shortly after that, Dr. Jesus Rosales-Ruiz invited Marilyn Gilbert to teach us instructional design. Marilyn has been wonderful in sharing her knowledge about him. Thinking about these events, I was very lucky to know him through people who understand his work very well. Though I could not meet him in person, I was just at the right place for me to know Tom Gilbert.

Now, my life cannot talk without Tom Gilbert. I wrote my doctoral dissertation, based on his book, under Dr. Rosales-Ruiz. It is an inquiry about the "ideal library." The performance matrix, an integration of four leisurely theorems, gave the library and information sciences a new way to look at what an ideal library would be. The performance matrix does not limit its application area or even the size of the target because it is always about engineering performance, especially worthy performance.

I just remembered a connection with Tom. It is an artistic sense. When I was invited to Marilyn's house, there was a big picture hanging at the entrance. I was very shocked, because it is by an artist from my hometown in Japan. He is not an artist that everyone knows. According to Marilyn, Tom brought it home from a pawnshop one day and thought a Chinese person had painted it. Well, close enough.

# Danny Langdon

*Originator of the Language of Work Model to Human Performance Technology*

## ✌ Tom Gilbert ✌

After more than forty years of practice, it's easy for me to reflect upon the men and women who influenced my own research and development in Human Performance Technology. The word *performance* itself automatically brings to mind Tom Gilbert. The father of engineering human performance, Tom has been cited time and time again for his leading work on mathetics and many other contributions. But his influence goes way beyond the obvious. Tom challenged many of us model-makers to be better and better, and was not hesitant with a sharp tongue for those who didn't measure up to what was already known as best practice. In Tom's way of thinking, you'd better know your stuff, whether as a researcher or practitioner! If you didn't, then you really didn't understand the foundation of our technology—performance itself—and probably didn't belong in the field. Thanks, Tom, for your quintessential work in our field and your influence on the way we practice.

# Joe Layng

*Co-founder and Senior Scientist, Headsprout*

## ◈ Remembering Tom ◈

Tom Gilbert is recognized as one of the great pioneers in helping to improve human performance and as a founder of Human Performance Technology as a discipline. His brilliance and flashes of insight are legendary. What seems at times to be overlooked, is his remarkable contribution to the field of instructional design.

I was first introduced to Tom Gilbert by Susan Markle. I was attending I believe an NSPI convention (now ISPI), and Sue asked if I would like to meet him. I was quite thrilled. Tom had already had quite an impact on the instructional design I was doing at that time. Sue had provided an introduction to two very important contributions of Tom's in her 1969 book *Good Frames and Bad*. The first was the concept of operant span. Many instructional designers had for the most part embraced the idea that programs had to be composed of small steps. What Tom had made clear, however, was that a program should not be made up of small steps, but the largest achievable step, that is, the learner's operant span. Sometimes this step might be

 Human Incompetence

simply to write a single word after reading a single sentence, or marking a matching picture after looking at another simple picture. At other times it might involve reading a chapter and calibrating a machine. It all depended on the entry repertoire of the learner and the task at hand.

The second contribution was mathetical instructional design. Ask someone about mathetics and if the person has heard of it at all, what one hears is basically the three fundamental characteristics of mathetical exercise design: demonstrate, prompt, and release. There is no doubt these principles have proven valuable over the years and are reflected in many successful programs including Direct Instruction (where the terms *model, lead, test* are used). However, mathetics as Tom envisioned was much more than that. For him, it was indeed a technology of education of which exercise design was but one part. For an educational effort to be truly mathetical in its approach it must also have a prescription, that is, a description of what the mastery performance must be, what Tom called the "synthetic repertoire." It must also have other key features, including the development of the Domain Theory, the process of extracting essential elements of the subject matter relevant to the prescription, and characterization, an analysis of the behavior properties of the prescribed repertoires. It was this analytical framework that brought the real power (and work) to mathetics.

For a program to be successful the thinking and analysis prior to actual program design has probably more to do with its success than any adherence to one exercise design philosophy or another. What Tom failed to mention, however, was how much damn work is involved!

As the years passed, I found myself frequently speaking with him at conventions and eventually he would visit me in Chicago. I particularly remember a visit in the early 1990s. We were watching one of our classes taught at the Personalized Curriculum Institute I had founded at Malcolm X College. A young woman in her late twenties was trying to learn to read beyond the second grade level she demonstrated upon entering the program. She was trying to read the word *street*. Every time she would say the word, however, she would pronounce it, "screet." The teacher would model the correct pronunciation and the student would repeat it, but it always came out "screet." Tom asked the teacher if he could try to teach the young woman to say street. He first asked the young woman if she could say "treat." She could. He then asked if she could make the sound "ssss." She could. He then told her to say "ssssss" pause a bit and say "treat." She could do this easily. He then had her reduce the pause with each pronunciation and within a few minutes she was happily pronouncing street as street. What Tom showed the teacher was that continued modeling and practice was not always the best solution. What was important was to look carefully at the properties involved in the subject matter and the performance and begin there. A little analysis could save a lot of teaching time, and using what was already in the repertoire of the learner could make all the difference. This was amply demonstrated in Tom and Marilyn Gilbert's great little book *Thinking Metric*, published in 1992.

This lesson, though obvious, was not lost on me. When it came time to write an online program that teaches children to read, we began with an analysis of the essential component and composite repertoires that needed to be established. We did not base our approach on the current orthodoxy about how decoding should be taught or how sounds should be presented or pronounced, but instead we focused on the accomplishments required given the entry repertoires of the children. We paid close attention to the operant span of our young learners and used mediators (different types of what might be called "memory aids"), as Tom advocated, where we could. In designing the first online comprehension program that actually teaches children how to comprehend, mediators form the core of several key teaching activities. Headsprout's reading programs would not have been the same, nor nearly as effective, without Tom. Even today he continues to touch learners all over the world, from Memphis, Tennessee to Port Elizabeth, South Africa, helping to teach over five hundred thousand (and growing) children how to read and comprehend.

# Maria E. Malott

## *CEO, Association for Behavior Analysis International*

 The Legacy of Thomas Gilbert: A Personal Account

I was fortunate to be accepted into the Behavior Analysis Program at Western Michigan University (WMU) to do my graduate work back in 1982. One of my first regular classes in the graduate program, taught by Norman Peterson, used Gilbert's *Human Competence*, which in my eyes, then and now, is a masterpiece.

I did not speak English then, and I had to stumble through the language in my first classes trying to understand reading assignments, not to mention the lectures and discussions. For the first time in my life, and to the disbelief of anyone who knew me, I got up at 4:00 a.m. every day, and, with a dictionary in hand, would take about an hour to digest each paragraph of my required reading. It was a painful experience indeed, until I came across *Human Competence*. At that juncture, I would look forward to my early mornings to continue to absorb this book, which transformed my professional and personal life in very profound ways. I still have my all-marked-up 1978 edition in my library as a guarded

 Human Incompetence

treasure. I would call my friends in Venezuela and tell them what Thomas Gilbert was teaching me. I would tell them how he provided answers to the many questions I had from my undergraduate experiences and from my first systems job before I came to WMU.

What was this excitement about? *Human Competence* taught me four lessons that have been ingrained in all my work: human capital potential, the analysis of performance problems, the perspective from different vantage points, and engineering performance systems.

## Human Capital Potential

Gilbert introduced the Potential for Improved Performance (PIP). Rather than assuming that we are born with a genetic combination that allows us to be exemplary, standard, or mediocre, PIP demonstrates that we all have the potential to be exemplary. We can study exemplary performance and compare it to anyone's performance and objectively measure the difference. Furthermore, we can figure out how to reduce the gap between them. "Yes, we can all be exemplars," I told my long-distance friends. Then I would try to learn what to do and how to do it from whoever I admired (including the great Thomas Gilbert himself). How could I learn from them? Opportunity for all was (and still is) very exciting to me!

## Analysis of Performance Problems

Coming from a culture where individuals are blamed for all systems' problems, Gilbert's concept of the Behavioral Engineering Model presents a wonderful contrast. No, it is not the individual's fault. We can narrow down performance problems to an objective analysis of the system and individual repertoires. Instead of blaming, we can analyze the causes of performance problems and find half of them are due to systems' problems, like lack of information, tools, or incentives. In cases in which the problem lay in individual repertoires, we can objectively analyze if the reason is lack of knowledge, skills, or the responsiveness to incentives, given the individual's behavioral history.

## Perspective from Different Vantage Points

The Performance Matrix is another invaluable concept because it acknowledges the importance of studying a system from multiple vantage points, from the philosophical and cultural to the logistic. To make a difference in a system, it is critical to not only examine the big picture, but also the administrative details as well.

## Engineering Performance Systems

With all said and done, Gilbert provides an integral perspective of how to engineer systems through which we can accomplish what the system was intended to achieve—starting with the impact in the culture and ending with the arrangement of the environment. This enables the work of individuals in the system to be consistent and aligned with the main aims of the system. Like architects design bridges and engineers build them, we can plan and manage human performance systems.

Gilbert not only influenced me directly, but also my mentors, Norman Peterson, Dale Brethower, and Richard Malott,

 Human Incompetence

whose teachings complimented and provided strengths to his concepts. In my practicum experience in Richard Malott's classes, I attempted to apply all the concepts from Human Capital. My master's thesis literally consisted of the application of *Human Competence* to the analysis of a higher education system. My doctoral dissertation, under the supervision of Dale Brethower, was on systems engineering and management—again, consistent with Gilbert's teachings.

Since I first read *Human Competence*, I have been trying to use the principles contained in it. Subsequently, I had many opportunities to interact with Gilbert personally and found all of them inspiring. Throughout my career, I have attempted to combine systems analysis with the analysis of individual behavior within systems in a great variety of businesses such as retail, manufacturing and service; government and private institutions; small and large organizations including a number of Fortune 500 businesses; and for-profit and non-profit organizations, including the management of the Association for Behavior Analysis International. In the process, I have taught dozens of corporate executives to appreciate the power of engineering human performance systems. The book, *Paradox of Organizational Change* (Malott, M. E., 2003) is an application and an extension of what I learned from Gilbert and my mentors, who also respected his work.

Gilbert's legacy will continue to live in me, in all my work, and in those I touch with it.

## Stuart Margulies

### *Work colleague*

I was one of Tom's groupies, not in the sexual way, but in the sense of admiring his work and often talking to other admirers about what we had learned. Two of the half dozen most skilled practitioners in the edutech world had no formal training; they just hung around Tom and became very expert. All the big shots tried to learn from them.

Instructional technology in the sixties was just beginning and there was nowhere to turn if you wanted to develop skill. Tom's papers were gems which all professionals treasured. But he had much less influence with the general reading public than people of modest skill. That's because he wrote in a non-technical way which made him seem unimportant, and because he gave his work such idiosyncratic names as *mathetics*. He wasn't a member of the establishment probably because he would not participate in low-quality, high-prestige conferences. Besides, I think everyone was afraid that Tom would make fun of them after their talks.

Rote learning was a topic tragically central to many experimental

 Human Incompetence

studies in those years. People who had to learn the names of all the bones and nerves of the body were taught in a way which simply didn't work. And the people who did these studies and formulated these useless techniques were very jealous of their methodology. I would have been scared to death to criticize them because they controlled access to grants and publication. The use of mnemonic devices was simply excluded from consideration, and besides it wasn't clear how to use mnemonics in a practical way. If anyone but Tom faced the problem of teaching what number corresponded to the color codes of the resistor, the instructional package would have been thirty or forty pages long, would have taken hours, and the reader would forget it the next day. Tom's paper on learning color codes of resistors was a revelation because he used mnemonics; "one brown penny" taught ten color codes, was very brief, taught with no fuss and the students never forgot. And many of us changed our whole approach to teaching names, regulations, and other rote material after reading this little program. I don't know whether it was Tom's brains or his balls that changed the field since he had to do it well and his approach was not proper.

Speaking of Tom's balls. When I worked for the federal agency Job Corps, I was unable to hire Tom as a consultant. He was openly living with his future wife who was not yet divorced from her husband. Tom was convicted of adultery (the last person convicted of this crime). The judge told him it was a "manly crime" and fined him fifty dollars but this ludicrous event prevented us from hiring him, although he was perhaps the person who could have helped us most.

His papers were diverse, too many to talk about here. One of his papers emphasized the difference between an employee not doing things well because he needed more training, versus not caring to do things well because he needed incentives. Perhaps that small paper should be on every list of the most important papers ever written in instructional technology.

If a seminal thinker is one who changes how people look at problems and how they try to solve them, Tom was one of the most important seminal thinkers of our era. We are all in debt to him.

# John McKee, Ph.D.

*Founder and Executive Director, Institute for Social and Educational Research*

*Founder and Chairman of the Board, Pace Learning Systems, Inc.*

## 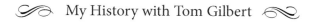 My History with Tom Gilbert

I first met Tom Gilbert in 1946 as a first-year graduate student in the Department of Psychology at the University of Tennessee. He was friendly and courteous but had nothing personal to say or observe. He entered graduate school along with two friends, all graduates of the University of South Carolina.

From the start, Tom seemed to be contentious and argumentative, but not in any belligerent manner. He questioned many assumptions of conventional psychology, perhaps seeking some solid principles and theories of psychology that he could accept.

Over time, Tom mellowed out and we became good friends. I accepted much of the clinical psychology that was being taught at the time, while Tom remained doubtful and somewhat contentious.

We graduated together and I recall that Tom's mother came to our graduation ceremony. Mrs. Gilbert was a businesswoman,

## Confessions of a Psychologist

operating one or more laundries, as I recall. It was a hot day as we three walked the campus. Finally, Mrs. Gilbert turned to us and said, "Well, I suppose you boys are proud of your degrees, but I look upon a degree in psychology as half common sense and half nonsense." With that she bade us goodbye so she could return to the real world of work and laundry. Tom was obviously embarrassed and made no comment and neither did I. His mother had embarrassed him and rather shocked me.

A number of years later, when Tom and I were together, I said, "Tom, do you remember when we were walking with your mother after our graduation, and she said that psychology was half common sense and half nonsense?" "Yes," said Tom, not embarrassed in the least now. "Mama was right!"

A few years later, I brought up the scene again and Tom had something different to say: "Mama was wrong. Psychology is *all* nonsense." We both had a good laugh over that.

Over the years, Tom remained a dear friend and we visited each other often. I learned a lot from Tom's questioning attitude and always respected it, since through it, I learned a lot and was able to develop applications of learning principles that were fruitful and respected by colleagues and business associates. I credit Tom then and now with the guidance, insight, and help in the application of psychology, applications I have used for solutions to instruction and learning problems, and for my work in mental health and corrections, as well as business and industry. I appreciate the great influence Tom had in my life and work.

# Byron Menides

*Vice President and CFO, Global Buyer's Networks, Ltd.*

Thomas F. Gilbert

**Background:**

I learned about Tom Gilbert from a lead article in a 1960 edition of *Fortune* magazine. The title of the article was "Programmed Instruction and Teaching Machine." It featured Crowder, scramble books, and programmed instruction. At the end of the article were comments from Thomas Gilbert who stated that a teaching machine was no more than a page-turner and the effectiveness of programmed instruction was determined by the quality and design of the program. I was impressed by his point of view, and his comments were, in my judgment, the highlight of the article.

At a friend's home in Wilton, Connecticut, I met Carl Sontheimer, who had founded two successful electronic companies and who read the same article in *Fortune* magazine. We both thought that education and training would be the prime focus in the 1960s as electronics was in the 1950s. We attended a conference on education and training held in Philadelphia where we spent most of our time at the exhibits area, talking to people representing their companies.

We were impressed with the Educational Design of Alabama's (EDA) exhibit and there we met Charles Slack representing the company.

We decided to meet the people involved with the EDA and met Tom Gilbert and the staff of EDA in the spring of 1961. I had an immediate liking of Tom who was articulate, friendly, personably, witty, and knowledgeable about effective methods of training. Carl and I had formed a company called Philotech, which we suggested would be the vehicle to acquire EDA after negotiating a fair price for the company. Tom discussed our offer with the key employees and the need to move to New York City. We negotiated terms, which were accepted by EDA. In June 1961, TOR Education, Inc. was incorporated and in July 1961, TOR exchanged 15 percent of its shares to acquire 100 percent of EDA. F.I. Rossmann & Co., an investment banking firm, raised $500,000 in a public offering in December 1961.

## Tom Gilbert's Impact of the Enterprise

Tom was the driving force in obtaining customers and developing programs to satisfy their needs. TOR obtained a contract with Harvard University for Timothy O'Leary to develop a word program, with American Airlines for a portion of the SABER operational systems, and with the U.S. Air Force for a special training program. Unfortunately the concept of programmed instruction was introduced during a time when there were very few early adopters and our key competitor Basic Systems was acquired by Xerox. Tom wrote and published the *Journal of Mathetics,* in part to generate business for TOR, but principally to introduce his

 Human Incompetence

unique system to training and educational "professionals," many of whom he thought were incompetent.

We were unable to generate any new customers. Members of TOR's board of directors, decided to wind-down TOR's activities, release all employees, and search for an acquisition to invest the firm's remaining funds.

A shareholder of TOR was also a major shareholder of International Accounting Society, Inc., a prestigious home study school founded in 1887 operating in Chicago. I met the CEO Clem McDonald and EVP Bill Rogers, who agreed to sell all of their controlling shares to TOR, which became a holding company. I moved to Chicago, and Tom and Marilyn move to New Jersey where he started Praxis. Its company mission was to improve a person's performance. Tom's method was based on observing a person exemplary performing a given task. He at times chose the second-best exemplary person. Based on the results of his primary research, he developed programs and guaranteed his customers that their employees would improve their performance on the task assigned. I was a member of the board of directors and was amazed how effective application of his systems were successful. During our five years, I remained in touch with Tom, and we met when we a had a meeting in NYC. While in Athens, Greece, at a chance meeting with Gilbert Granet, EVP of the Famous Artists School, at the hotel Grand Bretagne, he proposed to acquire TOR on very favorable terms. Ninety-eight percent of TOR shareholders approved the merger. After the merger, my wife Laura and I moved back to NYC and renewed our close friendship with the Gilberts.

Confessions of a Psychologist

## Personal

Before TOR acquired IAS, Laura and I lived in the Village and often met with the Gilberts for drinks, invigorating discussions, and dinner. On many Thursdays we would attend off-Broadway plays. We often met in their home, in New Jersey, our apartment in NYC, and their home in Dalton, Massachusetts. While the women would rap in another room, Tom and I would drink red wine and listen to records of Hank Williams. When our daughter was a student at the University of Virginia, to breakup the trip to Charlottesville, we would stay overnight at their home in New Jersey.

We would tell stories, and talk into the early morning. What wonderful times they were! Tom inherited the best attributes of Southern manners, character, and integrity. Tom would kid Laura who was a poet, that poems that didn't rhyme were not poems. He wrote some silly poems that rhymed and sent them to Laura to "see" how high she would jump. At their son's wedding, as usual we would stay with Tom and Marilyn overnight. Tom looked pale, weak, and frail. I feared that this would be the last time I would be with my dear friend.

# Laura Menides

*Member of the Board of Directors for the Worcester County Poetry Association*

## On Tom and Poetry

When I met Byron in July 1961, he and Carl Sontheimer had recently formed TOR Education, devoted to programmed learning, and were negotiating a merger with Educational Design of Alabama, EDA. They both spoke glowingly of EDA's brilliant young star, Tom Gilbert, whose practical, workable ideas would revolutionize the way that people could learn many tasks.

Of course, I was eager to meet Tom, which I did early that fall. Immediately, I understood Byron and Carl's enthusiasm. The Tom I met was indeed brilliant, quick-minded—and a real Southern charmer. Tall, very fair, thin and wiry, with a contagious laugh, Tom fascinated us all with his far-ranging knowledge, his fund of ideas and witticisms, as well as with his boundless energy and his eagerness to learn about us.

"Poetry? What kind of poetry?" Tom asked, when I mentioned that I was teaching poetry, and also writing and having my poems published. "I love poetry," he boomed,

Confessions of a Psychologist

and began to quote Robert Frost:

> *Whose woods these are I think I know;*
> *his house is in the village, though*
> *He will not see me stopping here*
> *to watch his woods fill up with snow,*
> *the darkest evening of the year.* . . .

I joined him, and together we happily recited the entire poem. Soon, however, the tone of our conversation changed. I mentioned that Frost's poetry was unusual in modern times, in that very often the lines of his poems were end-rhymed. My poetry is modern, I mentioned, and eschews end-rhyme. And I quoted Denise Levertov, who when she read in Massachusetts said, emphatically, "The modern world doesn't rhyme."

"Now, Laura," Tom began, in his most polite voice, "you know that's hogwash. If it doesn't rhyme, it's not poetry." Thus began a debate that Tom and I had each time we saw each other, and sometimes in between meetings, by letter. It was not harsh, but a loving, teasing debate. If he saw one of my poems, for example, he might say, "I enjoyed its sentiment and its power, but Laura, where's the rhyme? It's prose, not poetry."

And if occasionally Tom showed me one of his poems, invariably I'd want to praise his topic and his metrical skill, then point out that his rhymes were too conventional. "Why not think of less conventional rhymes, or re-write, eliminating the rhyme," I'd suggest. I remember quoting Ezra Pound, who insisted that modern poets should "make it new"—and that if they used rhyme at all, it should not

 Human Incompetence

be hackneyed like the "moon-June" variety, but rather fresh and new. Pound himself showed the way, for example, by using startling, bilingual rhymes, like "hair-Flaubert."

For years, Tom and I would bait each other in these friendly arguments, neither of us giving an inch, and each convinced we were right. I thought I had the last word. When Tom died in 1999, Marilyn had a lovely memorial service at their home in New Jersey, at which I read Seamus Heaney's description of a distinguished, intelligent, complex man, a description that fit Tom exactly. Heaney's piece was a poem—and unrhymed.

Before sitting down after the reading, however, I hesitated—and felt compelled to add that Tom no doubt, was looking down at me saying, "A lovely sentiment darlin,' but if it doesn't rhyme, it's not poetry."

On August 27, 2009, I met Marilyn for dinner in Seattle. It was an emotional moment for both of us because it was our first meeting since Tom died. We talked about our present situations and reminisced about the past; I sensed that Tom was enjoying our getting together.

# James S. Moore, Ph.D.

## *Friend, Colleague*

 **Thomas F. Gilbert and His Influence on My Professional and Political Activities**

Dr. Thomas F. Gilbert was an early and primary influence in my career as an educational psychologist and in my political activities as a candidate for Washington State Superintendent of Public Instruction in 1972.

Before Tom founded Human Performance Technology and wrote his book *Human Competence*, he was teaching and applying the work of B. F. Skinner in the foundational years of applied behavior analysis, instructional technology and teaching machines, and operant conditioning.

I became his graduate student assistant at the University of Georgia when Tom was the first psychologist in the nation to use Skinner's theories and research in paper-based technology that was to grow into computer assisted and computer-managed human learning. He applied and expanded B. F. Skinner's concept of teaching machines as a better way to blend traditional methods of knowledge-based teaching and systematically managing classroom behavior and performance. Tom was on the way to developing his Human

 Human Incompetence

Performance Technology (HTP) and his ABC Model (Antecedents-Behaviors-Consequences).

I learned to write linear-based instructional materials from Tom as a more efficient way than traditional textbooks. We chose this format over that of the so-called *scrambled book* and other formats because linear and sequential presentation of information seemed the more efficient way to apply the ABC Model. Research I initiated at The Human Resources Research Organiztion ( HumRRO) with Robert Mager, Paul Whitmore, and Charles Darby later showed that the "scrambled book" format created similar results.

At the University of Georgia Tom offered a graduate assistant position to me to study the operant behavior of animals in a laboratory setting. He was interested in showing that animal behavior and human behavior could be studied without the intrusive methods of chemical agents and mechanical implants, and that the findings from the ABC Model were just as valid as those from what was coming out of the work of physiological psychologists. At Georgia, Tom Gilbert taught that we did not need to surgically implant electrodes in animal brains, or blend water and drugs to influence and manipulate behavior.

My 1958 Master of Science thesis *Locomotor Operant Behavior in Male Albino Rats as a Function of Time* was supervised by Tom Gilbert who was my professor-advisor. Tom and I showed that animal behavior followed the principles of the ABC Model in that, over time, animals do not perform when their behavior does not lead to success, as when rats run a maze to locate food and eventually will not run the maze if food is no longer found. Tom loved to challenge physiological psychologists and my thesis study was another of his major efforts to create his Human Performance Technology and the ABC Model.

Tom Gilbert was mechanically inclined and built all of his equipment, including the maze mounted on a rocker arm and connected to an automatic counter that I used in our study of rats. After getting approval for me to conduct our study, Tom took sabbatical leave to work with B. F. Skinner at Harvard. He transferred his advisory tasks to Hudson Jost, the University of Georgia department chairman. I completed the masters degree under Jost, who later went to the chairmanship at Arizona State. Jost and Tom had an amiable relationship and Jost admired Tom's intellect and creativity.

After Tom returned in 1958, he found employment for me at HumRRO at Fort Bliss, Texas. His friend Joseph Hammock was the administrator and research director and had been one of Tom's colleagues. Hammock later was the department chairperson at the University of Georgia. Hammock assigned me to work with Bob Mager, Paul Whitmore, and Charles Darby in the TEXTTRUCT I program applying psychological learning theory and methods to the technical instruction and training of U.S. Army air defense personnel in the use of NIKE missile systems as anti-aircraft and anti-missile technology. Tom had taught me enough in the use of paper-based "teaching machines" so that I could assist the TEXTRUCT I team in applying programmed instruction. Tom supervised my work at a distance through Hammock, and made one on-site visit with me during my first year at Fort Bliss. I was following his advice and teaching, and learned skills and knowledge that would prove valuable in my later career as a school and educational psychologist.

I credit Tom Gilbert for most of the later developments in instructional technology. Directly and indirectly he influenced the work of Robert Mager and his colleagues at the Center for Effective Performance (CEP), and much of the initial work of Mager in his series of books on Criterion-Referenced Instruction (CRI). When I worked under Mager at HumRRO, I was asked to create paper-based materials for both the linear-sequential and the scrambled book formats to teach technical skills. Mager and his associates needed the expertise of someone trained in the Skinner-Gilbert theory and technology to expand what was being used at Fort Bliss, and to further enhance the research and development efforts of HumRRO to improve military applications.

When I left Fort Bliss in 1960 to do post-graduate work at the University of Washington in Seattle, the work of Skinner was being applied by Sidney Bijou and Donald Baer to research and development of materials and systems for children at the Child Development and Research Center (CDRC), most of whom were either handicapped or enrolled in the early childhood program.

In the two years I was on campus, I used the education and knowledge from my work with Tom Gilbert to make a decision to become a school-and-educational psychologist. The college of education was teaching traditional educational and teaching theory and methods. My life changed in those two years when I made a commitment to help the fields of education and psychology to recognize and apply the Skinner-Gilbert models.

I lost contact with Tom Gilbert after 1960 but I had the pleasure of applying what I learned from him in my work as a psychologist with the Seattle Public Schools. I traveled and studied various ways in which instructional technology was being applied in continuous-progress-education centers in New York City, Alabama, Florida, New Mexico, and California. All of them were struggling to use instructional technology and the Skinner-Gilbert models as well as the Mager-Madalyn Hunter models to see how the growing fields of programmed learning and instructional technology, including computer-based and computer-assisted technology, could be used to accomplish what Tom Gilbert called his Human Performance Technology (HPT) to improve student performance.

In 1972, as a candidate for Washington State Superintendent of Public Instruction, I made as my platform many of the aspects of what I had learned from Tom Gilbert, Skinner, and others in order to make more efficient use of public school funding, and to facilitate student performance with what Tom and Marilyn Gilbert called "The Science of Winning," based on their article on how University of Alabama coach Paul "Bear" Bryant built winning teams. I received 530,000 votes (46 percent of total), but lost to the candidate of the Washington Education Association-National Education Association, a union to which I belonged but which believed my opponent had more power to achieve legislative tax reform and full funding. I argued that instructional technology such as that of the Gilbert-Skinner-Mager models which I adopted as

 Human Incompetence

my own would make better use of current taxes and funding.

My memories of Tom Gilbert are very pleasant. He was exceptionally intelligent and enthusiastic about his research. He took me as a graduate assistant when he found my interests matched his own. I enjoyed his talks and ideas because they seemed much more meaningful for my concept of the field of psychology.

When Tom and Jost finally left the University of Georgia, Tom to go to Alabama, Jost to Arizona State, where he also died prematurely, their departures were a great loss for a growing department at Georgia. I regret the loss of Tom in 1994 and believe he would have been intellectually active and professionally productive for at least as long as was Skinner. They both were brilliant.

Tom's influence on my academic and intellectual growth and development was immense. If I had stayed at Georgia, most likely I would have been advised by him for my Ph.D. degree in the newly approved doctoral program. I needed income and left for HumRRO, but it was difficult to leave.

Tom was foremost in translating B. F. Skinner's theories and technologies into practical terms. I knew Tom to be equally intelligent and forward thinking to Skinner, and I am glad Tom was a friend of Skinner—and my friend also.

I miss him very much, but I have his book and works, as well as my keen memory of him to carry me on a little farther down the line. If there is a spiritual dimension, he and I will meet again. I believe it will happen.

# Tony Moore, CPT

## *Founder, Moore Performance Improvement, Inc.*

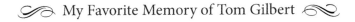
## My Favorite Memory of Tom Gilbert

When Dr. Ted Krein, senior manager at Ernst & Whinney, introduced me to the works of Tom Gilbert very early in my career, it changed my life. Ted's endorsement motivated me to read and reread *Human Competence* until I began to understand and appreciate the genius of Tom Gilbert.

A few years later, Joe Harless introduced me to Tom at an NSPI conference. I expressed my appreciation to Tom for all I had learned from his book. I later saw him sitting in the hotel lobby and asked him if he would mind signing my copy of his book, which I was rereading for the third or fourth time. Tom seemed genuinely delighted to be asked.

When he opened the cover and saw all the pages were covered with my comments, summaries, edits, and questions about what I was reading, he asked if I would mind sharing them with him.

We discussed my notes and his concepts for hours. Tom was eager to share his knowledge and was unbelievably

 Human Incompetence

patient and tolerant of my ignorance. For the rest of that evening, I was Tom's focal point, the fortunate recipient of his passion and, to quote Joe Harless, "his incomparable generosity concerning his ideas and his time."

When time caught up with us, a tired Tom Gilbert remembered that he hadn't yet signed my book, which was a Christmas present from my daughter, Ann. He put the following inscription immediately below hers: "For Tony, hope you turn out to be as competent as Ann. Love, Tom."

I'm still working on it, Tom.

# Richard M. O'Brien

*Professor of Psychology and a Core Faculty Member, Ph.D. Program in Combined Clinical and School Psychology, Hofstra University*

## Tom Gilbert: Visionary

I never worked with Tom Gilbert. I never studied with him. We were friends. We shared many of the same vices and the same interests. Chief among the latter and perhaps the former as well, was an addiction to baseball. I met Tom through Alyce Dickinson at a cocktail party at ABA many years ago. She introduced us by mentioning that at the time I was involved in a research project with the Major League Baseball Player Relations Committee. That bait was too enticing for Tom to resist. Holding out his hand and smiling he said "So you think you know baseball? Have you ever read Cook's statistical analysis of the best baseball players of all time? You can't possibly judge baseball talent if you haven't read Cook," he said.

The book he made reference to was Earnshaw Cook's *Percentage Baseball* (1966). I confessed that I had not read it but I would give my best infield of the players I had seen. I demurred that I couldn't speak for the real old timers. He took me up on it and I told him it would be Vic Power of the Athletics at first, Bill Mazeroski of the Pittsburgh Pirates

 Human Incompetence

at second, Brooks Robinson of the Orioles at third and, arguably, Roy McMillan of the Reds at shortstop. "Well you got three out of four; McMillan wasn't that good," he answered. We were glued to the bar talking baseball for the rest of the night and became fast friends for the rest of his life.

Tom and I did try to work together once. I had a project that I wanted to discuss with him so we met for lunch in the bar of the Summit Hotel near his home in New Jersey. We sat down at noon, left after six that evening, and never talked about the project. Although we did talk a little business and stop occasionally to admire a woman who walked by, we spent most of the afternoon talking baseball. We solved the problem of the designated hitter by concluding that it needs to vanish from the game. On the other hand, he failed to give me a satisfactory explanation for how he could come from South Carolina and be a Yankees fan of all things.

Tom Gilbert was a Southern "good old boy" whose writing style and language would have offended even a moderately insensitive feminist. Yet if you read *Human Competence*, you'll find in most of the hypothetical examples in which one manager does something correctly and the other botches it, the effective manager will be the female. This is a classic application of valuing accomplishment, effectively producing the desired outcome, over behavior, using the "politically correct" word to refer to a woman. To him, being concerned about word choice would have been just another example of the cult of behavior or valuing the act over the outcome. You get paid for what you produced not for what you did. If what you are measuring isn't still there after the employee goes home, you're measuring the wrong thing.

In my own experience, I had noticed that managers who had done the job themselves often micromanaged their employees when they became the supervisor. I observed that response but I didn't understand why they did it until I read *Human Competence*. They believed that because they had discovered the best way to do the job for them that was the best way for everybody to do the job. Such a conclusion violates the most basic tenets of logic but it can be observed in business every day.

Nonetheless, I did not immediately sign on to Tom's emphasis on outcomes over behavior. In fact, we had some discussions about this point as I tried to defend the parsimony of working directly with what the subject does rather than indirectly on an outcome that may or may not be traceable to what the subject actually did (O'Brien & Dickinson, 1982). I wasn't the only one to take that position at that time (See: Komaki, Collins & Thoene, 1980; Miller, 1978). Our disagreement on this point led me to try to answer the question empirically.

The family of one of my doctoral students, Bonnie Warren, owned a small factory that produced mattresses. It turns out the plant actually made conventional mattresses to be used on beds and couch mattresses referred to as "daveno mattresses," from the word *davenport,* for couches that folded out into beds. They had one assembly line so that they worked on only one type of mattress at a time. They

 Human Incompetence

would switch tasks throughout the day depending on what had been ordered. Baseline data showed that the line produced almost exactly the same quantity of each type of mattress, a little over eight per hour (Warren, 1982).

After a couple of trips to the plant, it was relatively easy to produce monitoring systems that independently measured behavior and outcomes. We measured behavior by having each of the men on the line record exactly what they were doing when someone was paged. That happened between ten and twenty times a day. To measure outcomes we had each employee on the assembly line record on the matress tag when he received a mattress. In this way, we knew exactly how long each employee had each mattress. That provided a measure of how long it took him to complete his step of the manufacturing process. Over a period of months, we spent half the month using the behavioral measure on the daveno mattresses and the outcome measure on the bed mattresses. On the second two weeks, the measurement systems were reversed. This system was in place over four phases (1. Baseline, 2. Monitoring, 3. Monitoring and Feedback, and 4. Monitoring, Feedback and Lottery Ticket Tangible Reinforcement for performance increases).

To my chagrin, the outcome-based measurement blew the doors off the behavioral system. In the monitoring phase, the behavioral measurement increased productivity by a little less than one piece per hour. The outcome system increased productivity by over eleven pieces per hour. It more than doubled productivity! While the initial jump in performance was not completely maintained, at no point, in any phase, with either task did the behavioral measurement outperform the outcome measurement.

Other than to like catfish, the importance of outcomes was the first thing that I learned from Tom. It was the first of many. I learned, for example, that if you continue to only attend to behavior you will miss the easy fix that comes from simply getting better equipment or distributing the equipment more effectively. I also saw demonstrations of the folly of rating behavior rather than counting it. I learned how easy it was to come up with the wrong measure as well as how impressed your consulting clients were when you pointed out the right ones. Tom pointed out the error of evaluating training departments based on number of classes held rather than the superior accomplishments of those who had been trained. Would I have realized as I once did with a client that evaluating the Maintenance Department based on the number of repairs they completed was self-defeating because that would mean that the ideal way for them to operate was to fix things poorly so they would quickly break again and have to repaired again? Without Tom Gilbert, I might not have.

When I shared Bonnie Warren's results with Tom, he was his usual modest self—once he had stopped laughing. Meanwhile I was searching for how I could have been wrong. What kind of a moderator could have produced this result?

This plant had not hired a worker for this assembly line in several years. Given the veteran status of the workers, we talked about what Gilbert called "deficiencies of execution" and "deficiencies of knowledge" or what Robert Mager

 Human Incompetence

(1970) called motivational deficits and skill deficits. These workers were not learning any new skills. Maybe measuring behavior is not so important if you are not trying to train new responses. Gilbert and I and many others had pointed out that you had to work with behavior when you were training new skills.

To save the operant model, one of my doctoral students and I mounted a study with the Hofstra University Varsity Baseball team. Hitting a round ball with a round bat when the ball travels over eighty miles per hour and doesn't go straight, has been called the most difficult feat in sports. Our baseball team wasn't doing it very well and hadn't for several years.

Susan Ross used teaching how to hit a baseball as the subject for her doctoral dissertation. She divided the team into three groups for a crossover design. After eight baseline games, she assigned one group to behavioral feedback and the second to outcome feedback. This was reversed after game fifteen. Group three received both kinds of feedback combined after game twenty-one. She provided feedback on appropriate batting behavior as suggested by the team coach as the behavioral feedback. The correct response was keeping the shoulders level during the swing as recommended by Ted Williams, the last man to have a .400 batting average. The outcome measure was how close to the sweet spot of the bat the ball had hit on each swing. Bats were chalked so the distance of the point of contact with the ball from the sweet spot could be measured after each swing. This is clearly an outcome of a behavior not a behavior in and of itself.

The results of the study were dramatic! Behavioral feedback turned out to be more effective than outcome feedback when there is a skill deficit. Further, the team which had lost all eight baseline games, won nine of the eighteen games during the intervention. During the eight baseline games, Hofstra scored a mean of 4.43 runs per game. For the eighteen games during treatment, the mean increased to 10.64 runs per game. The team also broke the school record for home runs in a season.

These results did not go unnoticed. The local press called me for a number of interviews including a lengthy one with the *New York Daily News*. In that interview, I attributed the success of what we had done to "teaching the batting techniques of Ted Williams using the performance improvement approach of Tom Gilbert." The next day I answered the phone in my office and heard a thick drawl say, "O'Brien, you made my life! You got me mentioned in the same sentence with Ted Williams!"

I surely didn't make Tom's life but he had a lot to do with making my career. I learned much of what I know about translating a functional analysis to a functioning organization from Tom Gilbert. He taught me how to advance an orthodox behavioral model by applying the same concepts to a bigger picture. I learned about PIPs, Acorn Tests, Accomplishments, Stakes, Values, and Performance Matrices. I have to thank Tom for the free use of all of these concepts. But, I also have to thank somebody else. I never heard Tom speak of his work without acknowledging the contributions of his wife Marilyn. In fact, as he talked, he usually pointed out that half of his ideas were actually Marilyn's and it was

 Human Incompetence

the better half.

To Tom, the worth of something was the ratio of its value to its costs. For me, what I got from Tom had inestimable value and it cost me nothing except some very well-spent afternoons. It was worth a great deal!

# Carol M. Panza

## *Management Consultant, CMP Associates*

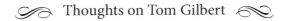 Thoughts on Tom Gilbert

One doesn't need to hear from folks like me, who actually knew Tom Gilbert, to figure out that he had a brilliant, original mind and that his work underlies most all of the truly effective tools, resources, and strategies for people (human) performance management and performance improvement. Why is this so important? Well, let's face it. People ARE the organization. You can have limitless financial resources, plant and equipment that are state of the art and at a scale to accommodate all levels of production, superior design for product and service offerings . . . well, you get the idea. If you have all the possible resources, but the organization's people are not suited for the jobs hey hold, skilled, engaged, organized and managed to collectively achieve customer-focused and organizationally desirable results, I contend that you won't be successful by any legitimate measure.

Tom articulated what, for me, is the fundamental concept regarding managing people, the concept of accomplishment, which allows us to define useful results-focused

Human Incompetence

position (job) requirements and measures, aligned with operating-process requirements and critical organization-level goals and objectives. But, that's not all, using accomplishment-based job descriptions and the logic that drives Tom's Performance Engineering Model, we have built practical, efficient, but most of all, effective, supervisory/management tools, resources and strategies. As a management consultant, practicing for the past thirty years, I believe that the concept of accomplishment has been largely overlooked and/or misunderstood. To me, however, it is one of the most powerful ideas that Tom developed. It is a truly essential insight, if your goals are achieving organizationally desirable results and organization success in their chosen marketplace.

Tom Gilbert's legacy to individual performance improvement professionals and organizations of all types is really all about results (accomplishment).

Accomplishments command far more agreement than activities or behaviors. By definition, they are measurable. Accomplishments remain useful for a long period of time, while at the same time supporting continuous improvement. That is, behaviors, techniques and tools can, and often should, change to achieve improved performance, while related outcomes or accomplishments will remain stable.

## A Slightly Odd Tribute

As was true, I'm sure, of many people, I knew of Tom Gilbert, through his work, some time before I ever met him.

## Confessions of a Psychologist

I actually met Tom for the first time in his home in New-Jersy. He graciously agreed to allow my friend and colleague, Randy James and I to come to his home. When we arrived, it was as if we had always known him. Tom was down to earth and fascinating to talk to, in his own quirky way. He was not the least bit pompous. He was eccentric, but charmingly so.

Randy reminded me recently about going out to lunch with Tom on that first visit. We took Tom's car because mine had only two seats. Tom asked Randy to drive and set expectations early on. "Pay no attention to the gauges on the dashboard. They don't work." So, on our way back from lunch, Randy was driving along unconcerned about an indication that the engine was overheating. We were laughing about what would otherwise have been some scary gauge readings, when Tom suddenly asked Randy to repeat the temperature reading. Apparently, that was the one gauge that did, in fact, function and it was off the charts (and not in a good way)! For me, besides walking back to Tom's house, it wasn't so much of an issue, since I lived in New Jersy and Tom didn't get excited. Randy, on the other hand, was ready to panic because he was scheduled to fly back to California that evening! In the end, Tom's car made it back to his house and Randy made it to the airport on time. More than twenty years later Randy and I still laugh about using Tom's car.

I'd like to contribute some further thoughts from the original version of a tribute I was asked to write for the special edition of *Human Competence* that was produced by ISPI after Tom passed away. Though the chart on the next page is not new, it is at the heart of my feelings about Tom as a

 Human Incompetence

professional, and is a fitting tribute to a truly gifted and genuinely unique man.

I call it the Guru Analysis Chart. I propose that Tom's actions and interactions with other professionals be studied as an example of a master performer guru. In this way, we can construct a really useful model of guru-ness.

| Guru Accomplishments | Tom/Master Performer | Average Gurus |
|---|---|---|
| Smiles and Laughter Produced | Incredible, if somewhat offbeat, sense of humor | Either no sense of humor or researches and memorizes anecdotes for all occasions |
| Ideas Disseminated and Implemented | An ability to explain complex concepts in a way that was not merely understandable, but more importantly, usable by us common folk | Explanations tend to make even simple concepts sound like they came from the Federal Register |
| People Reached/Touched | Open and approachable by most regular people especially those who really wanted to learn | Hold court on occasion, but are typically seen only with people of status |
| Technology Established | Conceptualized and articulated a technology of performance that is useful, enduring and able to be built on and extended/enriched | Profess ideas but those ideas are a sacred collection of statements only. No one else dare use them no less build on them, even were it possible. |

Useful, right? As Tom himself might suggest, let's contrast our master performer with the average guru.

For the first few years that I knew Tom, he referred to me as "the little blond girl." He couldn't remember my name, but he always remembered me. Tom's brilliance is a given and well known to many. What truly set him apart was his ability to engage others. Tom could stride right up to an individual or a crowd (cane in hand) and actually ignite a conversation that would involve, enlighten, and entertain all participants. He was one of a kind.

# Caroline Sdano, Ph.D.

## *Management Consultant and Writer*

### Tom Gilbert – An Evocation

When I remember Tom, I see him laughing. Unless you knew him well, you might not know that Tom had a finely honed sense of humor. I count myself as one of the lucky ones. I had both a personal and a professional friendship with Tom and with Marilyn for about the last twenty years of Tom's life.

It can't be easy to be a genius. Everyone else is so much slower than you are. They don't see the things that you do. Tom negotiated his way in this world of pygmies by being funny. This is a helpful posture for dealing with the inevitable dumbness of life's vagaries.

It is also a way of reminding you about your connection with someone, instantly. For example, his frequent greeting to me, "Caroline Sdano, a woman who doesn't know how to spell her own name." My rejoinder? "My daddy was a schoolteacher. We were too poor to buy a vowel."

 Human Incompetence

It is further a way to communicate with us lesser mortals. This was very important, given what I think was Tom's mission: to disseminate his attar of behavioral principles and the touchstones of teaching and learning he had developed. Who can forget how to distinguish between a training problem and a performance problem? "If someone put a gun to your head, could you do it if your life depended on it?"

There are thousands of ways that his notion of positive reinforcement has helped me, professionally and personally. Remembering to reward good behavior rather than punish bad behavior has helped me to manage clients and their aims better. It also helped me to raise my daughter Emily L. Mayer to be the lovely, accomplished adult that she is today, in my opinion.

Tom's search for ways to apply behavioral principles in everyday life could lead him on some quirky paths. He loved to explore all manner of what others might regard as gimcrackery and Rube Goldberg-esquery — just to see if they really worked. A bit of Frank B. Gilbreth showed in him when he elaborated on how to teach someone to play the clarinet while they wore a corset!

Tom's mnemonics could amount to the wisest aphorisms of the *Farmer's Almanac*. Thanks to Tom and Marilyn's *Thinking Metric*, I have no trouble with Celsius temperatures. I can see him exhorting, "Forty is fierce, thirty is thirsty, and twenty is plenty." How many times has it been useful to me to know that a nickel weighs five grams? Countless times.

For Tom was at bottom a teacher. Maybe that is why he is not as well known in the world at large as he should be (although he and Marilyn, all together, had more children than the Gilbreth's.)

I know he nurtured a wonderful sense of humor in the youngest children still at home when I knew Tom, son Rob and daughter Eve, and then, of course, granddaughter Georgia, who also lived with Marilyn and Tom. I remember going out in the middle of the night with a flashlight to scare the kids when they were camping in a tent in the backyard. They weren't frightened at all. It just increased the fun.

I am sure that the kids have many, many stories to illustrate how Tom lived with wit and wisdom. They all inherited his grin, his funniness, and his love of laughter. He had a most-useful approach for navigating life's highways and byways. It is a very valuable inheritance, one that will endure always.

# Julie Vargas, Ph.D.

*President, B.F. Skinner Foundation*
*Founding Editor, Behavior Analyst*

I first encountered Tom Gilbert's name in the early 1960s when I was working as a programmed instruction writer at the American Institutes of Research in Pittsburgh. Tom was well known among educational programmers for his innovative teaching of color codes for electrical resistors. Ten colors had to be paired with the numbers that indicated the ohms of resistance. Most programmers would have written sequences to teach each number one at a time, taking fifty or sixty frames to teach all the pairings. Tom used mnemonics like ONE BROWN penny, a WHITE cat has NINE lives, and ZERO; BLACK nothingness. Using these "mediators" he accomplished the teaching task in ten frames.

Tom ported procedures from the animal laboratory for more complex tasks. To teach a long series of steps to rats and pigeons, the last step is established first, then the next-to-last link in the chain and so on until you reach the first step. In human beings, then, to teach a task such as long division, the student is first presented with an almost-completed problem and asked to finish the last step. When that performance is firm, the instructional program backs up one step at a time until the student begins at the beginning.

In teaching this way, each completed step leads to a task the student can already do. That, along with correctly completing every problem, tends to be more reinforcing than encountering unfamiliar territory as work proceeds. Tom was known for promoting this "backward chaining," calling his system "mathetics." I remember a cartoon in which he was shown wearing a jacket backwards. He was shown as follows in William Deterline's 1967 comic depiction of programmed instruction.

"Gilbert described a systematic set of concepts and techniques that he called *mathetics* which sometimes made use of a technique called *chaining*. This confused many people because it seemed to teach things backwards."

In 1962 Tom published the *Journal of Mathetics*, but soon turned from formal education to working in business and industry. Being in education, I lost touch with his activities. I did meet him at conferences, however, and once visited his home in New Jersey when I was consulting nearby. He treated me graciously, always, as indeed he treated others.

## Kathleen Whiteside

*Founding Director, Partner, and Consultant,
Performance International*

*Co-editor: Intervention Resource Guide:
50 Performance Improvement Interventions*

### ～ Remembering Tom Gilbert ～

I attended my first ISPI (then NSPI) conference the same year that Tom published *Human Competence*, so I was way behind the curve of his friends, colleagues, and clients. Nonetheless, the very name of his book spoke deeply to me, for I was raised in a family that treasured competence over all other traits (including love, kindness, integrity…but that is another story for another time). It seemed to me, naive as I was at the time, that to have a system for engineering competence in people was only the best thing to have ever been invented. I worked my way through the book, absorbing ideas as best I could, and learning to apply his methodology at work and in my personal life.

It is also through my personal lens that I appreciate his invention of the field of mathetics. I had never done as well academically as I should have (given the native intelligence that was there) in large part because of learning disabilities, or poor organizational skills, or really poor memory. Whatever learning I mastered usually came from understanding the underlying logic inside the subject. When I learned about sequencing of learning blocks, I was ecstatic. And mathetics opened a whole new world to me. To this day, I remember "twenty is plenty" in terms of the centigrade scale. I have been able to tell whether I will like the day's temperature anywhere in the world.

So, although Tom has been gone for more than a decade now, and my major actual connection to him is through his wonderful wife, Marilyn, he is with me daily . . .an intellect that was a gift to the whole world.

# Acknowledgment

Performance Management Publications is indebted to Marilyn Gilbert for her generosity in sharing this special work for international publication. Marilyn now lives and works on Bainbridge Island, Washington. Her granddaughter Georgia Gilbert is still with her. Marilyn's writing course, Writing Solutions for Behavior Analysts, is now online at the University of North Texas. She continues to apply, in her words, "what Tom taught me to a subject matter I thought I already knew." Still very active in furthering her own work, Marilyn is currently writing a book on using the principles of performance and behavior analysis to teach writing at all levels. She plans to dedicate her book to Tom, because he would have loved it. "What a shame he wasn't here to help me when I got stuck!" as she so modestly says, since you have discovered in reading this autobiography, he would say that she so often helped him get "unstuck," with Marilyn being his first and foremost co-author and editor.

Photographs

# The Many Children in Tom's Life

*Marilyn and Charlie Ferster's children and Tom's and S.K. Dunn's child were part of Tom's life with Marilyn.*

*Andrea and Warren in Georgia, early 1960s*

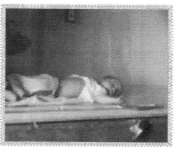

*Bill Ferster in his first year of life in his "baby box," 1956.*

*Warren and Sam Ferster in Georgia in the mid-1960s*

*Tom and Betty's children: Kathy, Sarah, Micha, Adam*

Confessions of a Psychologist

*Tom and Jessie (Tom and Katie Dunn's daughter) in mutual support, 1970s*

*Roby and Eve, Tom and Marilyn's children, 1970s*

*Tom's and Marilyn's son Robert (on right) as Oliver Twist in church play*

*Eve as Alice in Wonderland*

*Marilyn and Tom's daughter Eve*

*Marilyn and Tom's son Roby*

# Human Incompetence

*Roby & Robby Goodale, Friend of Tom's, whom Tom named his son after*

A photo I sent in the last batch is of Robby Goodale. He is with Robby Gilbert as an infant. He was a very good artist. We met him in Cuernavaca, Mexico, where he lived then. He shortly after that time moved to Union, Maine. They were very good friends almost immediately and until Robby died. Tom named our son Robby after him—although nowadays our Robby calls himself Roby, after Tom's 2nd middle name: Thomas Franklyn Roby Gilbert.

– ***Marilyn Gilbert***

Confessions of a Psychologist

Tom, Eve, and Georgia's first Christmas, December 1989

Marilyn with granddaughters, Margot Forster on right, Georgia Gilbert on left

Jessie and Tom at a wedding, 1990s

Eve, Georgia in carriage, Marilyn, Tom and Roby at State House in SC

At Georgia's dance recital

Marilyn and Tom on the South-Carolina beach, 1980s

*Revisting school in Dentonsville, SC, 1990s*

*Marilyn and Tom in the early 1990s*

*Marilyn and Tom at Georgia's dance recital, 1990s*

# References

# References

Cicerone, B. Improving Customer Satisfaction: A Lesson Learned By Finding The Reasons A Car Engine Does Not Start. *Performance Improvement,* 2009, Vol. 48, (August). Pages 5-9.

Cicerone, R.A. Keep Customers Coming Back - A Model And Job Aid For Creating Loyal Customers. *Performance Improvement,* 1998, Vol. 37, (July/August). Pages 57-62.

Cicerone, R.A., Sassaman, R., and Swinney, J. *Controlling the Behavior of Motor Vehicle Drivers:* A Nonverbal Application of Gilbert's Behavior Engineering Model. *Performance Improvement,* 2006, Vol. 45, (October). Pages 14-16. Performance And Instruction, 1986, Vol. 25 (September). Pages 3-6.

Cicerone, R.A., Sassaman, R., and Swinney, J. The Path to Improved Performance Starts with Theory: A Lesson Learned from Tom Gilbert. *Performance Improvement,* 2005, Vol. 44, (February). Pages 9–14.

Cook, E. *Percentage Baseball.* Boston: MIT Press, 1966.

Langdon, Danny *Aligning Performance: Improving People, Processes, and Organization,* John Wiley, 2000.

Dean, Peter J. Allow Me to Introduce…Thomas F. Gilbert. *Performance Improvement Quarterly,* 1992, 5 (3), 83-95.

Earle, R. Performance technology: A new perspective for

the public schools. *Performance Improvement Quarterly,* 1990, 3(4), 3-11.

Engelmann, S., & Carnine, D. W. *Theory of instruction: Principles and applications.* New York: Irvinston Publishers, 1982.

Gilbert, T. *Human Competence: Engineering Worthy Performance.* San Francisco: Pfeiffer (John Wiley & Sons, Inc.). © The International Society for Performance Improvement, 2007.

Gilbert, T. Mathetics: The technology of education. *Journal of Mathetics,* 1962a, 1, 7-74.

Gilbert, T. Mathetics II: The design of teaching exercises. *Journal of Mathetics,* 1962b, 1, 7-56.

Gilbert, T. F. Saying what a subject matter is. *Instructional Science,* 1976, 5, 29-53.

Gilbert, *T. F. Human Competence: Engineering Worthy Per formance.* New York: McGraw-Hill, 1978a

Gilbert, T. F. Guiding worthy performance. *Improving Human Performance Quarterly,*1978b, 7(3), 273-302,

Gilbert, T. F. A question of performance - part I: The PROBE model. *Training and Development Journal,* 1982a, September, 21-30.

Gilbert, T. F. A question of performance - part II: Applying the PROBE model. *Training and Development*

*Journal,* 1982b, October, 85-89.

Gilbert, T. F. *Measuring performance at work.* In Fishman, D. and Peterson, D. (Eds.), *Assessment for Decision,* Rutgers University Press 355-389, 1986.

Gilbert, T. F. Measuring the potential for performance improvement. *Training,* 1988a, July, 49-52.

Gilbert, T. F. and Gilbert, M.B. The science of winning. *Training,* August, 1988 34-40.

Gilbert, T. F. Thomas Gilbert: performance engineering. In G. Dixon (Ed.), *What Works at Work: Lessons from the Masters, Minneapolis*: Lakewood Books, 15-22, 1988b.

Gilbert, T. F., and Gilbert, M. B. Performance engineering: Making human productivity a science. *Performance and Instruction,* 1989, January 3.

Gilbert, T. F. *Maximizing the impact of training.* Workshop transcript, West Chester, Pennsylvania, 1990.

Gilbert T.F. and Gilbert M. B. *The considerable contributions of B F Skinner.* Unpublished paper, 1991a.

Gilbert, T. F. and Gilbert, M. B. *Taped interview with P.J. Dean and M. R. Dean,* Hampton, New Jersey, 1991b.

Gilbert, M. B. and Gilbert, T. F. What Skinner gave us. *Training,* 1991, September, 42-48.

Harless, J. (1989). Wasted behavior: A confession. *Training,* 1989, May, 35-38.

Johnson, K., & Street, E. M. *Direct Instruction. In the Encyclopedia of Educational Psychology,* 1, 240-243. Thousand Oaks, CA: Sage Publications, 2008.

Johnson, K., & Street, E. M. *The Morningside Model of Generative Instruction: An integration of research-based practices.* In D. J. Moran & R. Malott (Eds.). *Empirically supported educational methods.* St. Louis, MO: Elsevier Science/Academic Press, 2004.

Komaki, J., Collins, R. L. & Thoene, T. J. F. Behavioral measurement in business, industry & and government. *Behavioral Assessment, 1980,* 2, 103-123.

Mager, R. F. *Analyzing Problems, or You Really Oughta Wanta.* San Francisco:Fearon, 1970.

Markle, S. M. (1990). *Designs for Instructional Designers.* Champaign, IL: Stipes, 1990.

Markle, S. M. *Good Frames and Bad: A Grammar of Frame Writing* (2nd ed.). Hoboken, N.J.: John Wiley and Sons, Inc, 1969.

Miller, L. M. *Behavior Management: The New Science of Managing People at Work.* New York: Wiley, 1978.

O'Brien, R. M. & Dickinson, A. M. *Introduction to Industrial Behavior Modification.* In R. M. O'Brien, A. M. Dickinson, & M. P. Rosow (Eds.), *Industrial Behavior Modification: A Management Handbook* (pp. 7-34). Elmsford, NY: Pergamon Press, 1982.

Ross, S. A., O'Brien, R. M. & Mullins, W. C. *Two types of feedback to improve hitting in collegiate baseball. Paper presented at the meeting of the American Psychological Association, Toronto*, 1984, August.

Rummler, G. and Brache A. *Improving Performance: How to Manage the White Space on the Organization Chart.* (2nd edition) San Francisco: Jossey-Bass Publishers, 1995.

Skinner, B.F. Are theories of learning necessary? *Psychological Review*, 1950, 57, 193-216. Reprinted in Skinner, B.F. (1961). *Cumulative Record.* New York: Appleton-Century-Crofts.

Skinner, B.F. Behaviorism at Fifty. *Science,* 1963, 140, 951-958.

Warren, R. F. Improving productivity in an industrial setting: Differential effects of various response measures and secondary reinforcers. *Dissertational Abstracts International*: Section B. Sciences and Engineering, 1982, 43, 165.

Zemke, R. Tom Gilbert: The world is his laboratory. *Training*, 1984, December, 110-113.

Zigon, J. and Cicerone, R.A. Teaching Managers How To Improve Employee Performance. *Performance And Instruction* (published by the International Society for Performance Improvement, Silver Spring, MD), 1986, Vol. 25 (September). Pages 3-6.

# About ADI

Regardless of your industry or expertise, one thing remains constant: people power your business. At Aubrey Daniels International (ADI), we work closely with the world's leading organizations to accelerate their business performance by accelerating and sustaining the performance of the men and women whose efforts drive their success. We partner with our clients in a direct, practical, and sustainable way to get results faster and to increase organizational agility in today's unforgiving environment.

Founded in 1978, and headquartered in Atlanta, GA, we work with such diverse clients as Aflac, Duke Energy, Lafarge, Malt-O-Meal, M&T Bank, Medco, NASA, Roche Labs, Sears, and Tecnatom to systematically shape discretionary effort—where people consistently choose to do more than the minimum required. Our work with clients turns their strategy into action. We accomplish this not by adding new initiatives to their list, but by helping them make choices that are grounded in an ethical approach to people and business, by increasing effective and timely decision-making, and by establishing a culture of respect for each person's contribution, regardless of rank.

Whether at an individual, departmental, or organizational level, ADI provides tools and methodologies to help move people toward positive, results-driven accomplishments. ADI's products and services help anyone improve their business:

> **Assessments:** scalable, scientific analyses of systems, processes, structures, and practices, and their impact on individual and organizational performance
>
> **Coaching for Impact:** a behaviorally sound approach to coaching that maximizes individual contributions

**Surveys:** a complete suite of proprietary surveys to collect actionable feedback on individual and team performance, culture, safety, and other key drivers of business outcomes

**Certification:** ADI-endorsed mastery of client skills in the training, coaching, and implementation of our key products, processes, and/or technology

**Seminars:** a variety of engaging programs of practical tools and strategies for shaping individual and organizational success

**Scorecards & Incentive Pay:** an objective and results-focused alternative to traditional incentive pay systems

**Sustaining Lean-Sigma Gains:** a proactive and systematic process for managing safety that creates a culture of safe habits

**Behavior-Based Safety:** a proactive and systematic process for managing safety that creates a culture of safe habits

**Safety Leadership:** a behavioral approach to creating a high-performance safety culture through leadership action

**Expert Consulting:** custom, hands-on direction and support from seasoned behavioral science professionals in the design and execution of business-critical strategies and tactics

**Speakers:** accredited and celebrated thought leaders who can deliver the messages your organization needs on topics such as sustaining your gains, accelerating performance, and bringing out the best in others

aubreydaniels.com
aubreydanielsblog.com/
facebook.com/Aubrey.Daniels.International
twitter.com/aubreydaniels
youtube.com/aubreydaniels

Human Incompetence | Appendix

Performance Management Publications
# Additional Resources

**Safe by Accident? Take the Luck out of Safety**
Judy Agnew
Aubrey C. Daniels

**Removing Obstacles to Safety**
Judy Agnew
Gail Snyder

**Performance Management** *(4th edition)*
Aubrey C. Daniels
James E. Daniels

**Oops! 13 Management Practices that Waste Time and Money**
Aubrey C. Daniels

**Other People's Habits**
Aubrey C. Daniels

**Measure of a Leader**
Aubrey C. Daniels
James E. Daniels

**A Good Day's Work**
Alice Darnell Lattal
Ralph W. Clark

**Bringing Out the Best in People**
Aubrey C. Daniels

**You Can't Apologize to a Dawg!**
Tucker Childers

**Precision Selling**
Joseph S. Laipple

**The Sin of Wages!**
William B. Abernathy

For more titles and information call
**1.800.223.6191**
or visit our Web site
**www.PManagementPubs.com**

209

Appendix | **Register your Book**

# Register Your Book

Register your copy of *Human Incompetence* and receive exclusive reader benefits. Visit the Web site below and click on the "Register Your Book" link at the top of the page. Registration is free.

www.pmanagementpubs.com

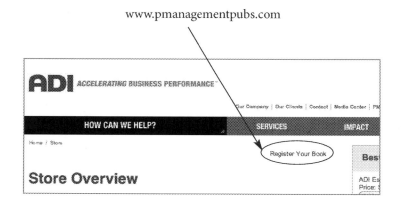